# A Student's Guide to Laplace Transforms

The Laplace transform is a useful mathematical tool encountered by students of physics, engineering, and applied mathematics, within a wide variety of important applications in mechanics, electronics, thermodynamics, and more. However, students often struggle with the rationale behind these transforms and the physical meaning of the transform results. Using the same approach that has proven highly popular in his other *Student's Guides*, Professor Fleisch addresses the topics that his students have found most troublesome, providing a detailed and accessible description of Laplace transforms and how they relate to Fourier and Z-transforms, written in plain language, and including numerous, fully worked examples. The book is accompanied by a website containing a rich set of freely available supporting materials, including interactive solutions for every problem in the text, and a series of podcasts in which the author explains the important concepts, equations, and graphs of every section of the book.

DANIEL FLEISCH is Emeritus Professor of Physics at Wittenberg University, where he specialises in electromagnetics and space physics. He is the author of five other books with the *Student's Guide* series, published by Cambridge University Press: *A Student's Guide to Maxwell's Equations* (2008); *A Student's Guide to Vectors and Tensors* (2011); *A Student's Guide to the Mathematics of Astronomy* (2013), *A Student's Guide to Waves* (2015), and *A Student's Guide to the Schrödinger Equation* (2020).

**Other books in the Student's Guide series:**

# A Student's Guide to Laplace Transforms

DANIEL FLEISCH
*Wittenberg University, Ohio*

CAMBRIDGE
UNIVERSITY PRESS

# CAMBRIDGE
## UNIVERSITY PRESS

University Printing House, Cambridge CB2 8BS, United Kingdom

One Liberty Plaza, 20th Floor, New York, NY 10006, USA

477 Williamstown Road, Port Melbourne, VIC 3207, Australia

314–321, 3rd Floor, Plot 3, Splendor Forum, Jasola District Centre,
New Delhi – 110025, India

103 Penang Road, #05–06/07, Visioncrest Commercial, Singapore 238467

Cambridge University Press is part of the University of Cambridge.

It furthers the University's mission by disseminating knowledge in the pursuit of
education, learning, and research at the highest international levels of excellence.

www.cambridge.org
Information on this title: www.cambridge.org/highereducation/isbn/9781009098496
DOI: 10.1017/9781009089531

First published 2022

*A catalogue record for this publication is available from the British Library.*

ISBN 978-1-009-09849-6 Hardback
ISBN 978-1-009-09629-4 Paperback

Additional resources for this publication at cambridge.org/fleisch-sglt.

# About this book

This edition of *A Student's Guide to Laplace Transforms* is supported by an extensive range of interactive digital resources, available via a companion website. These resources have been designed to support your learning and bring the textbook to life, supporting active learning and providing you with feedback. Please visit www.cambridge.org/fleisch-sglt to access this extra content.

We may update our Site from time to time, and may change or remove the content at any time. We do not guarantee that our Site, or any part of it, will always be available or be uninterrupted or error free. Access to our Site is permitted on a temporary and "as is" basis. We may suspend or change all or any part of our Site without notice. We will not be liable to you if for any reason our Site or the content is unavailable at any time, or for any period.

# Contents

# Preface

The purpose of this book is to help you build a foundation for understanding the Laplace transform and its relationship to the Fourier transform and the Z-transform. These transforms are useful in a wide variety of scientific and engineering applications, not only because they can be used to solve differential equations, but also because they provide an alternative perspective for extracting information from complex functions, signals, and data sequences. And although there are many conventional texts and websites that deal with integral and discrete-time transforms, the emphasis in those resources is often on the mechanics of taking the transform or its inverse. That is certainly important, but it's also important to develop an understanding of the rationale for the transformation process and the meaning of the result.

To help you develop that understanding, this book is written in plain language and is supported by a rich suite of freely available online materials. Those materials include complete, interactive solutions to every problem in the text, in-depth discussions of supplemental topics, and a series of video podcasts in which the author explains the key concepts, equations, and figures in every section of every chapter.

Like all the texts in Cambridge's *Student's Guide* series, this book is intended to serve as a supplement to the comprehensive texts that you may have encountered in your courses or as part of your continuing education. So although you will find plenty of examples of the forward and inverse Fourier, Laplace, and Z-transform in this book, those examples are designed to introduce fundamental concepts and techniques that will allow you to move on to intermediate and advanced treatments of these transforms.

If you have read any of my other Student's Guides, you're probably aware that I try hard to present challenging concepts in a way that's technically accurate but less intimidating than the presentation in many physics and engineering textbooks. In that effort, I'm inspired by the words of Thomas Sprat in his *History of the Royal Society*, in which he writes that the Royal Society encourages its members to use "a close, naked, natural way of speaking; positive expressions; clear senses; a native easiness: bringing all things as near the Mathematical plainness, as they can: and preferring the language of Artizans, Countrymen, and Merchants, before that, of Wits, or Scholars."

If that approach sounds about right, you may find this book helpful.

# Acknowledgments

This Student's Guide is the result of the efforts of many people, including Dr. Nick Gibbons, Dr. Simon Capelin, and the production team at Cambridge University Press. I thank them for their professionalism and support during the planning, writing, and production of this book.

I also owe thanks to Professor John Kraus of Ohio State, Professor Bill Gordon of Rice University, and Professor Bill Hunsinger of the University of Illinois, all of whom helped me understand the value of clear explanations and hard work. The support of Jill Gianola has been essential to my financial, physical, and emotional health.

# 1
# The Fourier and Laplace Transforms

The Laplace transform is a mathematical operation that converts a function from one domain to another. And why would you want to do that? As you'll see in this chapter, changing domains can be immensely helpful in extracting information from the mathematical functions and equations that describe the behavior of natural phenomena as well as mechanical and electrical systems. Specifically, when the Laplace transform operates on a function $f(t)$ that depends on the parameter $t$, the result of the operation is a function $F(s)$ that depends on the parameter $s$. You'll learn the meaning of those parameters as well as the details of the mathematical operation that is defined as the Laplace transform in this chapter, and you'll see why the Fourier transform can be considered to be a special case of the Laplace transform.

The first section of this chapter (Section 1.1) shows you the mathematical definition of the Laplace transform followed by explanations of phasors, spectra, and the Fourier Transform in Section 1.2. You can see how these transforms work in Section 1.3, and you can view transforms from the perspective of linear algebra and inner products in Section 1.4. The relationship between the Laplace frequency-domain function $F(s)$ and the Fourier frequency spectrum $F(\omega)$ is presented in Section 1.5, and inverse transforms are described in Section 1.6. As in every chapter, the final section (Section 1.7) contains a set of problems that you can use to check your understanding of the concepts and mathematical techniques presented in this chapter. Full, interactive solutions to every problem are freely available on the book's website.

## 1.1 Definition of the Laplace Transform

This section is designed to help you understand the answers to the questions "What is the Laplace transform?," and "What does it mean?" As stated above, the Laplace transform is a mathematical operation that converts a function of one domain into a function of a different domain; recall that the domain of a function is the set of all possible values of the input for that function. The domains relevant to the Laplace transform are usually called the "$t$" domain and the "$s$" domain; in most applications of the Laplace transform the variable $t$ represents time and the variable $s$ represents a complex type of frequency, as described below. The Laplace transform is an integral transform, which means that the process of transforming a function $f(t)$ from the $t$-domain into a function $F(s)$ in the $s$-domain involves an integral:

$$F(s) = \mathcal{L}[f(t)] = \int_{-\infty}^{+\infty} f(t)e^{-st}dt. \qquad (1.1)$$

So what does this equation tell you? It tells you how to find the $s$-domain function $F(s)$ that is the Laplace transform of the time-domain function $f(t)$. In the center portion of this equation, the expression $\mathcal{L}[f(t)]$ represents the Laplace transform as a "Laplace transform operator" ($\mathcal{L}$) that takes in the time-domain function $f(t)$, performs a series of mathematical operations on that function, and produces the $s$-domain function $F(s)$. Those operations are shown in the right portion of the equation to be multiplication of $f(t)$ by the complex exponential function $e^{-st}$ and integration of the product over time. The reasons for these operations are fully explained below.

You should be aware that Eq. 1.1, in which the integration is performed over all time, from $t = -\infty$ to $t = +\infty$, is the bilateral (also called the "two-sided") version of the Laplace transform. In many practical applications, particularly those involving initial-value problems and "causal" systems (for which the output at any time depends only on inputs from earlier times), you're likely to see the Laplace transform equation written as

$$F(s) = \int_{0^-}^{+\infty} f(t)e^{-st}dt, \qquad (1.2)$$

in which the lower limit of integration is set to zero (actually $0^-$, which is the instant just before time $t = 0$, as explained below) rather than $-\infty$. This is called the unilateral or "one-sided" version of the Laplace transform, and it is the form of the Laplace transform most often used in applications such as those described in Chapter 4.

If this is the first time you've encountered a zero with a minus-sign superscript ($0^-$), don't worry; the meaning of $0^-$ and why it's used as the lower limit of the unilateral Laplace transform are not hard to understand. The value of $0^-$ is defined by the equation

$$0^- = \lim_{\epsilon \to 0} (0 - \epsilon),$$

in which $\epsilon$ is a vanishingly small increment. Put into words, when applied to time, $0^-$ represents the time just before (that is, on the negative side or "just to the left" of) time $t = 0$.

Likewise, $0^+$ is defined by the equation

$$0^+ = \lim_{\epsilon \to 0} (0 + \epsilon),$$

which means that $0^+$ is just after (that is, on the positive side or "just to the right" of) time $t = 0$.

And here is the relevance of this to the unilateral Laplace transform: if you integrate a function using an integral with $0^-$ as the lower limit and any positive number as the upper limit, the value of the function at time $t = 0$ contributes to the integration process. But if the lower limit is $0^+$ and the upper limit is positive, the value of the function at $t = 0$ does not contribute to the integration. Since several applications of the unilateral Laplace transform involve functions with values that change significantly when their argument equals zero, integrating such functions using $0^-$ as the lower limit ensures that the value at $t = 0$ contributes to the result.[1]

The Laplace transform equation may look a bit daunting at first glance, but like many equations in physics and engineering, it becomes comprehensible when you take the time to consider the meaning of each of its terms. To do that, a good place to start is to make sure that you understand the answer to the question that confounds many students even after they've learned to use the Laplace transform to solve a variety of problems. That question is "What exactly does the parameter $s$ in the Laplace transform represent?".

As mentioned above, in most applications of the Laplace transform in physics, applied mathematics, and engineering, the real variable $t$ in the function $f(t)$ represents time and the complex variable $s$ in the function $F(s)$ is a generalized frequency that encompasses both a rate of decay (or growth) as well as a frequency of oscillation. Like any complex number, the $s$-parameter in the Laplace transform can be written as the sum of a real part and an imaginary part:

$$s = \sigma + i\omega, \tag{1.3}$$

in which $\sigma$ represents the real part and $\omega$ represents the imaginary part of the complex variable $s$, and $i$ represents the imaginary unit $i = \sqrt{-1}$.

---

[1] A complete discussion of the importance of $0^-$ to the Laplace transform can be found in Kent H. Lundberg, Haynes R. Miller, and David L. Trumper "Initial Conditions, Generalized Functions, and the Laplace Transform," http://math.mit.edu/~hrm/papers/lmt.pdf.

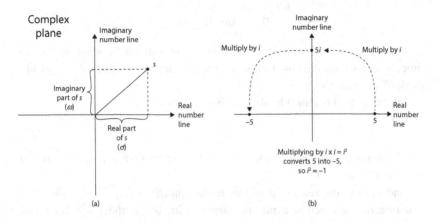

Figure 1.1 (a) Laplace parameter $s$ on the complex plane, and (b) why $i^2 = -1$.

To understand the nature of a complex number, it's helpful to graphically represent the real part and the imaginary part of a complex number on two different number lines, as shown in the "complex plane" diagram in Fig. 1.1a. As you can see, in the complex plane the imaginary number line is drawn perpendicular to the real number line.

So what does this have to do with the imaginary unit $i$? Consider this: if you multiply a number on the real number line, such as 5, by the imaginary unit $i$, that real number 5 becomes the imaginary number $5i$ because it's now on the imaginary number line, as shown in Fig. 1.1b. And if you then multiply again by $i$, you get $5i \times i = -5$. So if multiplying by $i \times i$ converts the number 5 into $-5$, then $i^2$ must equal $-1$, which means that $i$ must equal $\sqrt{-1}$. Since squaring any real number can't result in a negative value, it's understandable that $i$ has come to be called the imaginary unit.

The imaginary part of $s$ (that is, $\omega$) is the same angular frequency that you may have encountered in physics and mathematics courses, which means that $\omega$ represents the rate of angle change, with dimensions of angle per unit time and SI units of radians per second. Since radians are dimensionless (being the ratio of an arc length to a radius), the units of radians per second (rad/sec) are equivalent to units of 1/seconds (1/sec), and this means that the result of multiplying the angular frequency $\omega$ by time ($t$) is dimensionless.[2] That is reassuring since $st = (\sigma + i\omega)t$ appears in the exponent of the term $e^{st}$ in the Laplace transform.

---

[2] Note that the abbreviation "sec" is used for "seconds" rather than the standard "s"; this is done throughout this book in order to avoid confusion with the Laplace generalized frequency parameter.

Since $\omega$ has dimensions of 1/time and SI units of 1/sec, Eq. 1.3 makes sense only if $\sigma$ and $s$ also have dimensions of 1/time and SI units of 1/sec. For this reason, you can think of $s$ as a "generalized frequency" or "complex frequency," in which the imaginary part is the angular frequency $\omega$. The meaning of the real part ($\sigma$) of the Laplace parameter $s$, and why it's reasonable to call it a type of frequency, is explained in Section 1.5.

If you're wondering about the dimensions of the Laplace transform output $F(s)$, note that the time-domain function $f(t)$ can represent any quantity that changes over time, which could be voltage, force, field strength, pressure, or many others. But you know that $F(s)$ is the integral of $f(t)$ (multiplied by the dimensionless quantity $e^{-st}$) over time, so the dimensions of $F(s)$ must be those of $f(t)$ multiplied by time. Thus if $f(t)$ has dimension of voltage (SI units of volts), the Laplace transform $F(s)$ has dimensions of volts multiplied by time (SI units of volts-seconds). But if $f(t)$ has dimensions of force (SI units of newtons), then $F(s)$ has dimensions of force multiplied by time (SI units of newtons-seconds).

## 1.2 Phasors and Frequency Spectra

Before getting into the reasons for using a complex frequency parameter in a Laplace transform, it's helpful to consider the product of the angular frequency $\omega$ with the imaginary unit $i = \sqrt{-1}$ and time $t$ in the exponential function $e^{i\omega t}$. This produces a type of spinning arrow sometimes called a "phasor," a contraction of the words "phased vector." As time passes, the phasor represented by $e^{i\omega t}$ rotates in the anticlockwise direction about the origin of the complex plane with angular frequency $\omega$. As you can see in the upper left portion of Fig. 1.2, the angle (in radians) that this phasor makes with the positive real axis at any time $t$ is given by the product $\omega t$, so the larger the value of $\omega$, the faster the phasor rotates. And if $\omega$ is negative, the phasor $e^{i\omega t}$ rotates in the clockwise direction.[3]

You should also remember the relationship between the exponential function $e^{i\omega t}$ and the functions $\sin(\omega t)$ and $\cos(\omega t)$ is given by Euler's relation:

$$e^{\pm i\omega t} = \cos(\omega t) \pm i \sin(\omega t), \tag{1.4}$$

---

[3] Some students, thinking of frequency as some number of cycles per second, wonder "How can anything rotate a *negative* number of cycles per second?" That question is answered by thinking of frequency components in terms of phasors that can rotate either clockwise (negative $\omega$) or anticlockwise (positive $\omega$).

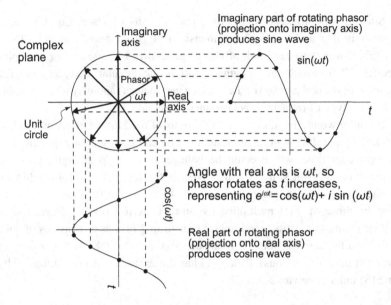

Figure 1.2  Rotating phasor relation to sine and cosine functions.

which is also illustrated in Fig. 1.2. As shown in the figure, the projection of the rotating phasor onto the real axis over time traces out a cosine function, and the phasor's projection onto the imaginary axis traces out a sine function. Adding a time axis as the third dimension of the complex-plane graph of a rotating phasor results in the three-dimensional plot shown in Fig. 1.3, in which the solid line represents the path of the tip of the $e^{i\omega t}$ phasor over time.

So the exponential function $e^{i\omega t}$ is complex, with the real part equal to $\cos(\omega t)$ and the imaginary part equal to $\sin(\omega t)$. Also helpful are the inverse Euler relations, which tell you that the cosine and sine functions can be represented by the combination of two counter-rotating phasors ($e^{i\omega t}$ and $e^{-i\omega t}$):

$$\cos(\omega t) = \frac{e^{i\omega t} + e^{-i\omega t}}{2} \tag{1.5}$$

and

$$\sin(\omega t) = \frac{e^{i\omega t} - e^{-i\omega t}}{2i}. \tag{1.6}$$

(If you'd like to see how these counter-rotating phasors add up to give the cosine and sine functions, check out the problems at the end of this chapter and the online solutions).

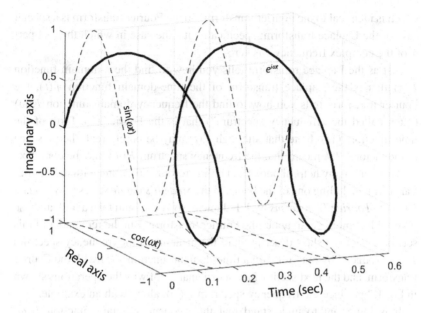

Figure 1.3 The path of the tip of the rotating phasor $e^{i\omega t}$ over time with $\omega = 20$ rad/sec.

This means that the real angular frequency $\omega$ does a perfectly good job of representing a sinusoidally varying function when inserted into the exponential function $e^{i\omega t}$. But that may cause you to wonder "Why bother making a complex frequency $s = \sigma + i\omega$?"

To understand the answer to that question, it's helpful to begin by making sure you understand the Fourier transform, which is a slightly simpler special case of the Laplace transform. The equation for the Fourier transform looks like this:

$$F(\omega) = \int_{-\infty}^{+\infty} f(t)e^{-i\omega t}\,dt. \tag{1.7}$$

Comparing this equation to the equation for the bilateral Laplace transform (Eq. 1.1), you can see that these equations become identical if you set the real part $\sigma$ of the Laplace complex frequency $s$ to zero, which makes $s = \sigma + i\omega = 0 + i\omega$. That makes the bilateral Laplace transform look like this:

$$F(s) = \int_{-\infty}^{+\infty} f(t)e^{-(\sigma+i\omega)t}\,dt = \int_{-\infty}^{+\infty} f(t)e^{-i\omega t}\,dt, \tag{1.8}$$

which is identical to the Fourier transform. So the Fourier transform is a special case of the Laplace transform; specifically, it's the case in which the real part $\sigma$ of the complex frequency $s$ is zero.[4]

Just as the Laplace transform tells you how to find the $s$-domain function $F(s)$ that is the Laplace transform of the time-domain function $f(t)$, the Fourier transform tells you how to find the frequency-domain function $F(\omega)$ (often called the "frequency spectrum") that is the Fourier transform of the time function $f(t)$. Note that although $f(t)$ may be purely real, the presence of the factor $e^{-i\omega t}$ means that the frequency spectrum $F(\omega)$ may be complex.

There are many helpful books and online resources dealing with the Fourier transform, including one of the books in the *Student's Guide* series (*A Student's Guide to Fourier Transforms* by J. F. James), so you should take a look at those if you'd like more detail about the Fourier transform. But the remainder of this section contains a short description of the meaning of the frequency spectrum $F(\omega)$ produced by operating on a time-domain function $f(t)$ with the Fourier transform, and the next section has an explanation of why the operations shown in Eq. 1.7 produce the frequency spectrum $F(\omega)$, along with an example.

It is important to understand that the frequency-domain function $F(\omega)$ contains the same information as the time-domain function $f(t)$, but in many cases the frequency-domain representation may be much more readily interpreted. That is because the frequency spectrum $F(\omega)$ represents a time-changing quantity (such as a voltage, wave amplitude, or field strength) not as a series of values at different points in time, but rather as a series of sinusoidal "frequency components" that add together to produce the signal or waveform represented by the time-domain function $f(t)$. Specifically, for a real time-domain function $f(t)$, at any angular frequency $\omega$, the real part of $F(\omega)$ tells you how much $\cos(\omega t)$ is present in the mix, and the imaginary part of $F(\omega)$ tells you how much $\sin(\omega t)$ is present. And although the function $f(t)$ that represents the changes in a quantity over time can look quite complicated when graphed, that behavior may be produced by a mixture of a reasonably small number of sinusoidal frequency components. An example of that is shown in Fig. 1.4, in which $f(t)$ represents a time-varying quantity and $F(\omega)$, the Fourier transform of $f(t)$, is the corresponding frequency spectrum (for simplicity, only the positive-frequency portion of the spectrum is shown). As you can see, trying to determine the frequency content using the graph of $f(t)$

---

[4] Note that this does not mean that you can find the Fourier transform $F(\omega)$ of any function simply by substituting $s = i\omega$ into the result of the Laplace transform $F(s)$ – for that to work, the region of convergence of the Laplace transform must include the $\sigma = 0$ axis in the complex plane, as discussed in Section 1.5.

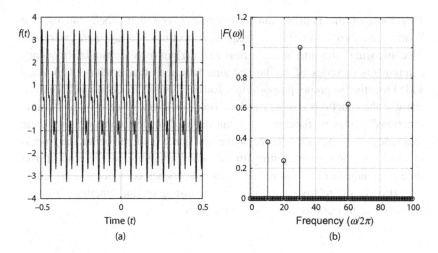

Figure 1.4 (a) Time-domain function $f(t)$ and (b) magnitude of frequency-domain function $F(\omega)$.

shown in Fig. 1.4a would be quite difficult. But the graph of the magnitude of the frequency-domain function $|F(\omega)|$ in Fig. 1.4b makes it immediately obvious that there are four frequency components present in this waveform. The frequency of each of those four components is given by its position along the horizontal axis; those frequencies are 10, 20, 30, and 60 cycles per second (Hz) in this case. The height of each peak indicates the "amount" of each frequency component in the mix that makes up $f(t)$; in this case those relative amounts are approximately 0.38 at 10 Hz, 0.23 at 20 Hz, 1.0 at 30 Hz, and 0.62 at 60 Hz.

So that's why it is often worth the effort to calculate the frequency spectrum $F(\omega)$ by taking the Fourier transform of $f(t)$. But how exactly does multiplying $f(t)$ by the complex exponential $e^{-i\omega t}$ and integrating the product over time accomplish that?

You can get a sense of how the Fourier transform works by remembering that multiplying $f(t)$ by $e^{-i\omega t}$ is just like multiplying $f(t)$ by $\cos(\omega t)$ and by $-i\sin\omega t$, so the Fourier transform can be written like this:

$$F(\omega) = \int_{-\infty}^{+\infty} f(t)e^{-i\omega t}\,dt$$

$$= \int_{-\infty}^{+\infty} f(t)\cos(\omega t)\,dt - i\int_{-\infty}^{+\infty} f(t)\sin(\omega t)\,dt.$$

This form of the Fourier transform reminds you that when you use the Fourier transform to generate the frequency spectrum $F(\omega)$, you are essentially "decomposing" the time-domain function $f(t)$ into its sinusoidal frequency components (a series of cosine and sine functions that produce $f(t)$ when added together in proper proportions). To accomplish that decomposition, you can use the functions $\cos(\omega t)$ and $\sin(\omega t)$ for every value of $\omega$ as "testing functions" – that is, functions that can be used to determine whether these frequency components are present in the function $f(t)$. Even better, the process of multiplication by these testing functions and integration of the product over time is an indicator of "how much" of each frequency component is present in $f(t)$ (that is, the relative amplitude of each cosine or sine function). You can see how that process works in the next section.

## 1.3 How These Transforms Work

To understand how the process of decomposing a time-domain function into its sinusoidal frequency components works, consider the case in which the time-domain function $f(t)$ is simply a cosine function with angular frequency $\omega_1$. Of course, for this single-frequency case you can see by inspection of a graph of $f(t)$ that this function contains only one frequency component, a cosine function with angular frequency $\omega_1$, but the decomposition process works in the same way when a cosine or sine function is buried among many other components with different frequencies and amplitudes.

Figures 1.5 and 1.6 show what happens when you multiply $f(t) = \cos(\omega_1 t)$ by cosine and sine functions (the real and imaginary parts of $e^{-i\omega t}$). In Fig. 1.5,

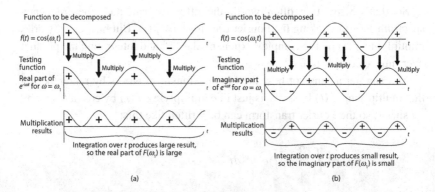

Figure 1.5 Multiplying $f(t) = \cos(\omega_1 t)$ by (a) the real portion and (b) the imaginary portion of $e^{-i\omega t}$ when $\omega = \omega_1$.

the frequency $\omega$ of the testing function exactly matches the frequency $\omega_1$ of the time function $f(t)$. As you can see in Fig. 1.5a, the peaks and valleys of the cosine testing function line up in time with the peaks and valleys of $f(t)$, so the result of multiplying these two functions together is always positive, and adding the products together over time results in a large real number (in fact, it will be infinitely large if $\omega$ precisely matches $\omega_1$ and you integrate over all time). So for $f(t) = \cos(\omega_1 t)$, the multiply-and-accumulate process results in a frequency spectrum $F(\omega)$ with a large real value at frequency $\omega_1$.

Now look at Fig. 1.5b. The same process of multiplication and summation gives a very different result when the testing function is a (negative) sine function, even though the frequency of that sine function matches the frequency $(\omega_1)$ of $f(t)$. Since the peaks and valleys of the sine function are offset from the peaks and valleys of $f(t)$, the result of multiplying these functions point by point varies from positive to negative, and summing the products over time and multiplying by $-i$ results in a small imaginary number (which will be zero if you integrate over all time, or over an integer number of cycles if $f(t)$ is periodic). That means that for $f(t) = \cos(\omega_1 t)$, the multiply-and-accumulate process results in a frequency spectrum $F(\omega)$ with a small imaginary value at frequency $\omega_1$.

So that's what happens when the frequency of the testing functions matches a frequency component of $f(t)$. What happens when you use testing functions with frequencies that don't match one of the frequency components of $f(t)$?

You can see two examples of that in Fig. 1.6. In Fig. 1.6a, the frequency of the cosine testing function is twice the value of the sole frequency component of $f(t)$ (that is, $\omega = 2\omega_1$). In this case, some of the peaks of the testing function line up with the peaks of $f(t)$, but others line up with valleys. That means that

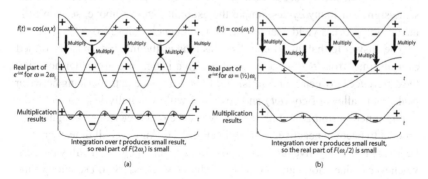

Figure 1.6 Multiplying $f(t) = \cos(\omega_1 t)$ by the real portion of $e^{-i\omega t}$ when (a) $\omega = 2\omega_1$ and (b) $\omega = \omega_1/2$.

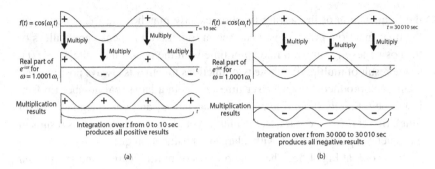

Figure 1.7 Multiplying $f(t) = \cos(\omega_1 t)$ by the real portion of $e^{-i\omega t}$ with $\omega = 1.0001\omega_1$ over time (a) 0 to 10 seconds, and (b) 30 000 to 30 010 seconds.

the multiplication results are both positive and negative, and integration over time produces a small number. The same is true for the sine testing function, so both the real and imaginary parts of the spectrum $F(\omega)$ are small in this case. And if the frequency of the testing function is smaller than the frequency component of $f(t)$, the result is also small, as shown in Fig. 1.6b for the cosine testing function with $\omega = \omega_1/2$.

Seeing the results of these multiplications for frequencies such as $\omega = 2\omega_1$ and $\omega = \omega_1/2$, some students wonder why this "testing" process works when the frequency of the testing function is only slightly different from a frequency component of $f(t)$ (rather than double or half of $\omega_1$ as shown in Fig. 1.6). For example, if $\omega = 1.001\omega_1$, the spacing between the peaks and valleys of the testing function is only slightly smaller than the spacing between the peaks and valleys of the function $f(t) = \cos(\omega_1 t)$. In such cases, the point-by-point multiplication process between the functions produces results with the same sign over many cycles, unlike the examples shown above, so won't the integration process yield a large result for frequencies close to $\omega_1$?

The answer to that question is "No," as long as the integration is performed over all time, from $t = -\infty$ to $t = +\infty$. That is because no matter how tiny the difference between $\omega$ and $\omega_1$ is, the slight difference in the spacing between peaks and valleys of $\cos(\omega t)$ and $\cos(\omega_1 t)$ will eventually lead to the peaks of one function lining up with the valleys of the other, causing the sign of the product to become negative, as shown in Fig. 1.7b. And over all time, there will be just as much negative product as positive, and the integration will yield zero whenever $\omega$ does not equal $\omega_1$. This is why it is necessary to use a long time window if you hope to distinguish frequencies that are close to one another (that is, if you want your frequency spectrum to have "fine resolution").

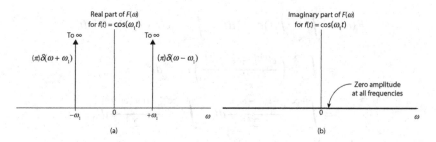

Figure 1.8 (a) Real and (b) imaginary parts of frequency spectrum $F(\omega)$ produced by the Fourier transform of $f(t) = \cos(\omega_1 t)$.

So it's through this point-by-point multiplication and integration process that the Fourier transform produces a result of zero for $F(\omega)$ whenever $f(t)$ is multiplied by a sine or cosine with a frequency that doesn't match one of the frequency components that make up $f(t)$ and the product is integrated over all time. And when the frequency of the multiplying function does precisely match one of the frequencies present in $f(t)$, the peaks and valleys will all align, the products will all be positive, and the result of integrating over all time will be infinitely large.

This infinitely tall and infinitely narrow result of applying the Fourier transform to the time-domain function $f(t) = \cos(\omega_1 t)$ can be represented graphically by a Dirac delta function located at the angular frequency $\omega$ that matches the frequency component $\omega_1$ in $f(t)$; this delta function is written as $\delta(\omega - \omega_1)$. And since the cosine function is an even function, which means that $\cos(\omega t) = \cos(-\omega t)$, the same result occurs at $-\omega$. Hence the real part of the frequency spectrum $F(\omega)$ of $f(t) = \cos(\omega_1 t)$ has two spikes, one at $\omega = +\omega_1$ and the other at $\omega = -\omega_1$, as shown in Fig. 1.8a. The imaginary part of the frequency spectrum $F(\omega)$ of a pure cosine wave has zero amplitude at all frequencies, as you can see in Fig. 1.8b, since no sine components are needed to make up a pure cosine function.

As indicated in this figure, the spectrum $F(\omega)$ is written as

$$F(\omega) = \pi[\delta(\omega + \omega_1) + \delta(\omega - \omega_1)], \qquad (1.9)$$

which often causes students to wonder about the origin and meaning of the factor of $\pi$ in front of each delta function. After all, isn't $\pi$ times infinity just as large as infinity?

To understand where that factor comes from and what it means, remember the inverse Euler relation for the cosine function (Eq. 1.5), which means the Fourier transform for $f(t) = \cos(\omega_1 t)$ can be written as

$$F(\omega) = \int_{-\infty}^{+\infty} f(t)e^{-i\omega t}\,dt = \int_{-\infty}^{+\infty} [\cos(\omega_1 t)]e^{-i\omega t}\,dt$$

$$= \int_{-\infty}^{+\infty} \left(\frac{e^{i\omega_1 t} + e^{-i\omega_1 t}}{2}\right)e^{-i\omega t}\,dt$$

$$= \int_{-\infty}^{+\infty} \frac{e^{-i(\omega-\omega_1)t}}{2}\,dt + \int_{-\infty}^{+\infty} \frac{e^{-i(\omega+\omega_1)t}}{2}\,dt.$$

These integrals can be replaced with the definition of the Dirac delta function:

$$\delta(\omega) = \frac{1}{2\pi}\int_{-\infty}^{+\infty} e^{-i\omega t}\,dt \qquad (1.10)$$

or

$$\delta(\omega + \omega_1) = \frac{1}{2\pi}\int_{-\infty}^{+\infty} e^{-i(\omega+\omega_1)t}\,dt,$$

$$\delta(\omega - \omega_1) = \frac{1}{2\pi}\int_{-\infty}^{+\infty} e^{-i(\omega-\omega_1)t}\,dt, \qquad (1.11)$$

which makes the Fourier transform of a pure cosine function

$$F(\omega) = \int_{-\infty}^{+\infty} \frac{e^{-i(\omega-\omega_1)t}}{2}\,dt + \int_{-\infty}^{+\infty} \frac{e^{-i(\omega+\omega_1)t}}{2}\,dt$$

$$= 2\pi\frac{\delta(\omega-\omega_1)}{2} + 2\pi\frac{\delta(\omega+\omega_1)}{2}$$

$$= \pi[\delta(\omega+\omega_1) + \delta(\omega-\omega_1)]. \qquad (1.12)$$

So that's where the factor of $\pi$ comes from, but to understand the meaning of a multiplicative term in front of a Dirac delta function, recall that the delta function is actually a distribution rather than a function. The utility of a distribution such as $\delta(x)$ is provided not by its numerical value at $x = 0$, but rather by its effect on other functions. The relevant effect in this case is produced by the "sifting" property of $\delta(x)$:

$$\int_{-\infty}^{+\infty} f(x)\delta(x)\,dx = f(0). \qquad (1.13)$$

When the delta function appears under an integral such as this, the meaning of a multiplicative term becomes clear:

$$\int_{-\infty}^{+\infty} f(x)[\pi\,\delta(x)]\,dx = \pi\int_{-\infty}^{+\infty} f(x)\delta(x)\,dx = \pi f(0). \qquad (1.14)$$

So in this case the delta function "sifts out" the value of the function at $x = 0$, and it is that value that gets multiplied by $\pi$.

Now consider the Fourier transform of the sum of two cosine functions with different amplitudes and different angular frequencies, such as $f(t) = A \cos(\omega_1 t) + B \cos(\omega_2 t)$. In that case, the Fourier transform results in a frequency spectrum given by

$$F(\omega) = \int_{-\infty}^{+\infty} \frac{A e^{-i(\omega-\omega_1)t}}{2} dt + \int_{-\infty}^{+\infty} \frac{A e^{-i(\omega+\omega_1)t}}{2} dt$$

$$+ \int_{-\infty}^{+\infty} \frac{B e^{-i(\omega-\omega_2)t}}{2} dt + \int_{-\infty}^{+\infty} \frac{B e^{-i(\omega+\omega_2)t}}{2} dt$$

$$= (\pi A)[\delta(\omega + \omega_1) + \delta(\omega - \omega_1)] + (\pi B)[\delta(\omega + \omega_2) + \delta(\omega - \omega_2)].$$

Thus the Fourier transform sorts the frequency components and tells you the amplitude of each, as you saw in the example with four frequency components shown in Fig. 1.4. A very similar analysis can be applied to the Fourier transform of a pure sine function, although the presence of the factor of $i$ in the denominator of the inverse Euler relation for the sine function (Eq. 1.6) and the sign difference between the exponential terms leads to a different result. Here is the Fourier transform of the pure sine function $f(t) = \sin(\omega_1 t)$:

$$F(\omega) = \int_{-\infty}^{+\infty} f(t) e^{-i\omega t} dt = \int_{-\infty}^{+\infty} (\sin(\omega_1 t)) e^{-i\omega t} dt$$

$$= \int_{-\infty}^{+\infty} \left( \frac{e^{i\omega_1 t} - e^{-i\omega_1 t}}{2i} \right) e^{-i\omega t} dt$$

$$= \int_{-\infty}^{+\infty} \frac{e^{-i(\omega-\omega_1)t}}{2i} dt - \int_{-\infty}^{+\infty} \frac{e^{-i(\omega+\omega_1)t}}{2i} dt$$

$$= (i\pi)[\delta(\omega + \omega_1) - \delta(\omega - \omega_1)], \tag{1.15}$$

in which the relation $1/i = -(i * i)/i = -i$ has been used. This purely imaginary frequency spectrum is shown in Fig. 1.9. In this case, the real part of the spectrum has zero amplitude everywhere, since no cosine components are needed to make up a pure sine wave.

The process of transforming the time-domain function $f(t)$ into the frequency-domain function $F(\omega)$ using the Fourier transform (that is, of decomposing a time function into its sinusoidal frequency components), is sometimes referred to as "Fourier analysis". The major steps in this process are shown in Fig. 1.10. And as you may have surmised, there's an inverse process that "goes the other way," meaning that it transforms the frequency-domain

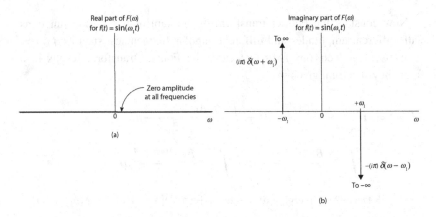

Figure 1.9 (a) Real and (b) imaginary parts of frequency spectrum $F(\omega)$ produced by the Fourier transform of $f(t) = \sin(\omega_1 t)$.

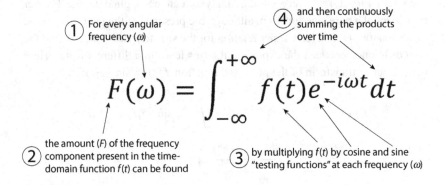

Figure 1.10 Fourier analysis.

function $F(\omega)$ into the time-domain function $f(t)$. That process is the inverse Fourier transform, which can be written as

$$f(t) = \frac{1}{2\pi} \int_{-\infty}^{\infty} F(\omega)e^{i\omega t}\,d\omega. \tag{1.16}$$

The most significant difference between the Fourier transform (sometimes called the "forward Fourier transform") and the inverse Fourier transform is that the time-domain function $f(t)$ and the frequency-domain function $F(\omega)$ have switched positions. But notice also that there is a constant factor of $1/(2\pi)$ in front of the inverse transform, the multiplying function is $e^{i\omega t}$ rather than $e^{-i\omega t}$, and the integration is performed over frequency rather than time.

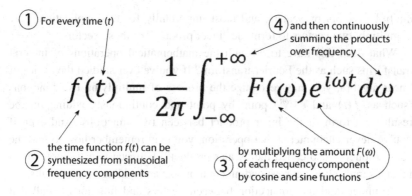

Figure 1.11 Fourier synthesis.

The reasons for these differences are explained in Section 1.6, but for now you should just be aware that the process of putting together sinusoidal functions to produce a time-domain function is accomplished with the inverse Fourier transform. That process is called "Fourier synthesis," and its major steps are shown in Fig. 1.11.

And just as the inverse Fourier transform converts the frequency-domain function $F(\omega)$ to the time-domain function $f(t)$, the inverse Laplace transform converts the generalized frequency-domain ($s$-domain) function $F(s)$ to the time-domain function $f(t)$. You can read more about inverse transforms in Section 1.6, and you can see the inverse Laplace transform in action in each of the examples in Chapter 2.

## 1.4 Transforms as Inner Products

Once you've achieved a reasonably good understanding of the Fourier transform and how it works, you may benefit from a somewhat different perspective on the process of finding the frequency spectrum of a time-domain signal. That perspective comes from linear algebra and the treatment of functions as objects that behave like vectors; a collection of these "generalized vectors" forms an abstract vector space. If you're not sure what that means, this may help: a space of abstract vectors is called a "space" because it's where all these objects can be found, "abstract" because this space is not the same as the physical universe we inhabit, and "vectors" because these objects obey the mathematical rules of vectors. Those rules tell you how to add two or more vectors together, how to multiply vectors by scalars, and most importantly for this application, how to multiply two vectors to form the "inner product" of those vectors.

What does this have to do with the mathematical operations of integral transforms such as the Fourier transform? If you've ever studied the basics of linear algebra, you may recognize the process of multiplying two functions (such as $f(t)$ and $e^{-i\omega t}$) point by point and continuously adding up the result as a form of the inner product between two functions. And even if you've never encountered that operation, you may remember how to find the "dot product" (also called the "scalar product") between two vectors, which involves multiplying corresponding components and adding up the results.

To understand the connection between vectors and functions, recall that vectors can be written or "expanded" as combinations of scalar coefficients and directional indicators called "basis vectors." So two vectors, $\vec{A}$ and $\vec{B}$, can be written as

$$\vec{A} = A_x \hat{\imath} + A_y \hat{\jmath} + A_z \hat{k}$$

and

$$\vec{B} = B_x \hat{\imath} + B_y \hat{\jmath} + B_z \hat{k}, \tag{1.17}$$

in which $A_x$, $A_y$, and $A_z$ represent the coefficients of vector $\vec{A}$, $B_x$, $B_y$, and $B_z$ represent the coefficients of vector $\vec{B}$, and $\hat{\imath}$, $\hat{\jmath}$, and $\hat{k}$ represent the Cartesian basis vectors pointing in the directions of the $x, y$, and $z$ axes, respectively. Written this way, each vector appears as a linear combination of basis vectors, with each basis vector weighted by the relevant coefficient. So the coefficients tell you "how much" of each basis vector is present in the vector.

Importantly, the same vector may be written using different basis vectors, so you could express vectors $\vec{A}$ and $\vec{B}$ using $2\hat{\imath}$, $2\hat{\jmath}$, and $2\hat{k}$ as basis vectors, for example, or using a completely different coordinate system such as spherical coordinates with basis vectors $\hat{r}$, $\hat{\theta}$, and $\hat{\phi}$. Of course, if you transform to a system with different basis vectors, the vector coefficients will generally change, because it may take different amounts of those new basis vectors to make up the vector. But expanding the vector in terms of different basis vectors does not change the vector itself, because the *combination* of coefficients and basis vectors will add up to give the same vector.

In the three-dimensional Cartesian coordinate system, the dot or scalar product between $\vec{A}$ and $\vec{B}$ can be written as

$$\vec{A} \circ \vec{B} = A_x B_x + A_y B_y + A_z B_z = |\vec{A}||\vec{B}|\cos\theta, \tag{1.18}$$

in which $|\vec{A}|$ and $|\vec{B}|$ represent the magnitudes (lengths) of vectors $\vec{A}$ and $\vec{B}$ and $\theta$ is the angle between $\vec{A}$ and $\vec{B}$.

One useful way to interpret the meaning of the dot product between two vectors is to consider the result of the dot product as an indication of how much one vector lies in the direction of another vector. In other words, the dot product can be used to answer the question "If I travel from the start to the end of one of the vectors, how far have I gone in the direction of the other vector?". If the two vectors lie in approximately the same direction, the answer will be larger than if the two vectors are nearly perpendicular. So the dot product is largest when the two vectors lie in the same direction ($\cos 0° = 1$), and the dot product is zero when the two vectors are perpendicular to one another ($\cos 90° = 0$).

The dot product is especially helpful when one or both of the two vectors is a basis vector. For example, the basis vectors $\hat{\imath}$, $\hat{\jmath}$, and $\hat{k}$ are frequently defined to be "orthonormal," meaning that each has unit magnitude (length of one) and that each is perpendicular to the other two. In that case, dot products between these basis vectors give these results:

$$\hat{\imath} \circ \hat{\imath} = \hat{\jmath} \circ \hat{\jmath} = \hat{k} \circ \hat{k} = 1$$
$$\hat{\imath} \circ \hat{\jmath} = \hat{\imath} \circ \hat{k} = \hat{\jmath} \circ \hat{k} = 0.$$

Now consider what happens when you form the dot product between one of the orthonormal basis vectors and a vector such as $\vec{A}$:

$$\hat{\imath} \circ \vec{A} = \hat{\imath} \circ (A_x\hat{\imath} + A_y\hat{\jmath} + A_z\hat{k}) = A_x(\hat{\imath} \circ \hat{\imath}) + A_y(\hat{\imath} \circ \hat{\jmath}) + A_z(\hat{\imath} \circ \hat{k})$$
$$= A_x(1) + A_y(0) + A_z(0) = A_x.$$

So if you want to find the $x$-component of vector $\vec{A}$, you can take the dot product between the unit vector in the $x$-direction ($\hat{\imath}$) and $\vec{A}$. And you can find the other components of $\vec{A}$ using the other orthonormal basis vectors:

$$A_x = \hat{\imath} \circ \vec{A} \qquad A_y = \hat{\jmath} \circ \vec{A} \qquad A_z = \hat{k} \circ \vec{A}. \qquad (1.19)$$

Note that if the basis vectors are not orthogonal, the dot products between different basis vectors will not be zero, and these simple relations will not work.

Here is the key to understanding the Fourier transform from the perspective of linear algebra: just as a vector can be decomposed into a weighted linear combination of basis vectors, so too can a function be decomposed into a set of "basis functions" that can be combined in the proper proportions to give the desired function. In the case of a time-domain function such as $f(t)$, the basis functions are an infinite set of delta functions, one at every value of time, and the coefficients are the values of the function $f(t)$ at each time. When you perform a Fourier transform on the function $f(t)$, the result is the frequency spectrum $F(\omega)$, for which the basis functions are the sine and cosine functions at every angular frequency, and the coefficients are the values of the function $F(\omega)$ at each frequency. That is why you sometimes see the Fourier transform described as an operation that changes from the "time basis" to the "frequency basis".

To extend the concept of functions as abstract vectors a bit further, recall that whenever one vector is a positive multiple of another, such as vector $\vec{A}$ and vector $5\vec{A}$, those two vectors point in the same direction (although the length of vector $5\vec{A}$ is five times the length of vector $\vec{A}$). Likewise, a function that is a multiple of another function, such as $\cos(\omega t)$ and $5\cos(\omega t)$, can be thought of as "lying in the same direction" in the abstract vector space that contains the functions. And just as two vectors are perpendicular when their dot product (formed by multiplying corresponding components and adding up the results) equals zero, two functions are considered to be orthogonal if their inner product (formed by multiplying the functions point by point and continuously summing the results) equals zero. So just as there are "lengths" and "directions" in the vector space containing vectors such as $\vec{A}$ and $5\vec{A}$, there are generalized "lengths" and "directions" in the abstract vector space that contains functions such as $\sin(\omega t)$ and $\cos(\omega t)$.

What makes the complex exponential function $e^{i\omega t}$ (and the sinusoidal functions that it encompasses) especially useful as basis functions is that the functions $\sin(\omega t)$ and $\cos(\omega t)$ are orthogonal to one another (they have zero inner product when integrated over all time or over any integral number of cycles, as explained in the previous section), and the functions $\cos(\omega_1 t)$ and $\cos(\omega_2 t)$ are orthogonal whenever $\omega_1$ does not equal $\omega_2$. This is the foundation of the "orthogonality relations" between sinusoidal functions, and it means that the $\sin(\omega t)$ and $\cos(\omega t)$ functions form a set of orthogonal basis functions.

Just as you can use the dot product to find the amount of each orthogonal basis vector present in a vector (as in Eq. 1.19), you can also use the inner product to find the amount of each orthogonal basis function in any function. The inner product between two functions $g(t)$ and $f(t)$, often written as $\langle g, f \rangle$, is defined as

$$\langle g, f \rangle \equiv \int_{-\infty}^{+\infty} g^*(t) f(t) dt, \qquad (1.20)$$

in which the asterisk in $g^*(t)$ indicates the complex conjugate.[5]

---

[5] You can find an explanation of why the complex conjugate is needed, along with more details about functions as abstract vectors in the "Vectors and Functions" document on this book's website.

Applying the inner product to complex exponential basis functions $e^{i\omega t}$ and the time-domain function $f(t)$ looks like this:

$$\langle e^{i\omega t}, f(t) \rangle = \int_{-\infty}^{+\infty} (e^{i\omega t})^* f(t)dt = \int_{-\infty}^{+\infty} f(t)e^{-i\omega t}dt \qquad (1.21)$$

and comparison to Eq. 1.7 shows that this inner product equation is identical to the equation of the Fourier transform. Hence you can consider the frequency spectrum $F(\omega)$ to be the inner product between orthogonal sine and cosine basis functions and $f(t)$.

Whether you think of the Fourier transform as multiplying the time-domain function $f(t)$ by sinusoids and integrating over time or as taking the inner product between functions, the result is the same: $F(\omega)$ tells you the "amount" of each frequency component present in the function $f(t)$. The Laplace transform output $F(s)$ also tells you about the frequency content of a time-domain function, but the presence of the real part $\sigma$ in the complex frequency parameter $s$ permits the Laplace transform to exist in some cases in which the Fourier transform does not. The next section of this chapter explains why that is true and describes some important differences between the frequency spectrum $F(\omega)$ produced by the Fourier transform and the output $F(s)$ of the Laplace transform.

## 1.5 Relating Laplace $F(s)$ to Fourier $F(\omega)$

As described in Section 1.2, the equation for the bilateral Laplace transform is identical to the equation for the Fourier transform when the real part $\sigma$ of the complex frequency $s = \sigma + i\omega$ is zero, making the term $e^{-\sigma t}$ in the Laplace transform equal to one. But the real power of the Laplace transform comes about when $\sigma$ is not zero and $e^{-\sigma t}$ is not one. The goal of this section is to help you understand the meaning of $\sigma$ and role of the real exponential term $e^{-\sigma t}$ in allowing the Laplace transform to operate on time-domain functions $f(t)$ for which the Fourier transform does not exist.

An expanded version of the bilateral (two-sided) Laplace transform is shown in Fig. 1.12. Note that when $\sigma$ is not equal to zero, $f(t)$ is multiplied not only by $e^{-i\omega t}$ as in the Fourier transform, but also by $e^{-\sigma t}$.

By grouping the $e^{-\sigma t}$ term with $f(t)$, you can see that the Laplace transform operates in the same way as the Fourier transform, with an important difference: that operation is performed not on $f(t)$, but on $f(t)$ multiplied by the real exponential $e^{-\sigma t}$. Unlike the complex exponential $e^{-i\omega t}$, which

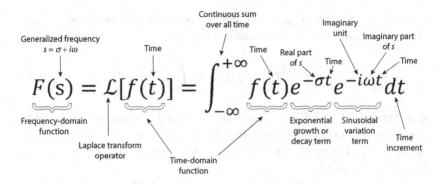

Figure 1.12 Bilateral Laplace transform equation.

oscillates sinusoidally over time, the real exponential $e^{-\sigma t}$ increases (if $\sigma$ is negative) or decreases (if $\sigma$ is positive) monotonically with increasing time.

So what is the benefit of multiplying $f(t)$ by the real exponential $e^{-\sigma t}$ before performing a transform? In some cases, the Fourier transform of a time-domain function doesn't exist because the integral in the Fourier transform doesn't converge – that is, the continuous sum of the product of $f(t)$ and the complex exponential $e^{-i\omega t}$ is infinite. For example, recall from the discussion in Section 1.3 that the integral in the Fourier transform of a pure cosine wave $f(t) = \cos(\omega_1 t)$ (which, by definition, goes on forever) gives an infinite result when the angular frequency $\omega$ of the testing function matches $\omega_1$. In that case, the frequency spectrum $F(\omega)$ could be written in closed form only by substituting the definition of an infinitely tall and infinitely thin distribution (the Dirac delta function) in place of the integral of a complex exponential over all time .

For many time-domain functions $f(t)$, the difficulty of nonconverging integrals in the transform can be avoided by multiplying $f(t)$ by a factor that decreases over time before operating with the Fourier transform (that is, before multiplying by the complex exponential $e^{-i\omega t}$ and integrating over time). The real exponential $e^{-\sigma t}$ is just such a factor, as long as $\sigma$ is a positive, real constant and you perform the integration over positive time only. This may seem like a severe restriction, but many practical applications of integral transforms involve causal functions of time, which means that $f(t) = 0$ for all values of $t$ less than zero. But even when you're dealing with noncausal functions that may grow large at increasingly negative values of time, you may still be able to achieve convergence by using negative values of $\sigma$, since the

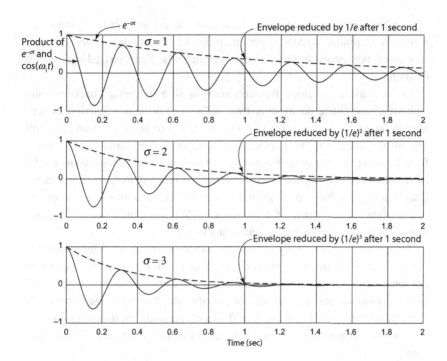

Figure 1.13 Product of exponential-decay term and $\cos(\omega_1 t)$ for $\sigma = 1$/sec, 2/sec, and 3/sec.

real exponential factor $e^{-\sigma t}$ decreases as time becomes more negative if $\sigma$ is also negative.

Of course, the Laplace transform integral will still yield an infinite result if the time-domain function $f(t)$ increases faster over time than the $e^{-\sigma t}$ factor decreases. And what determines how fast $e^{-\sigma t}$ decreases with time? The value of $\sigma$.

A closer look at the meaning of $\sigma$ makes it clear that $\sigma$ is indeed a type of frequency (recall from Section 1.1 that $\sigma$ has SI units of 1/seconds). To see that, consider the value of $e^{-\sigma t}$ if both $\sigma$ and $t$ equal one. At that instant, the factor $e^{-\sigma t} = e^{-(1)(1)} = \frac{1}{e} = 0.368$, which means the envelope of whatever you've multiplied by this factor, such as the $\cos(\omega_1 t)$ function in Fig. 1.13, has been reduced by 63.2% from its value at time $t = 0$. But if $\sigma = 2$, then at that same time ($t = 1$), the factor $e^{-\sigma t} = e^{-(2)(1)} = \frac{1}{e}\frac{1}{e} = 0.135$, which means the amplitude has decreased by 86.5%. And if $\sigma = 3$, then after one second $e^{-\sigma t} = \left(\frac{1}{e}\right)^3 = 0.05$, a decrease in amplitude of 95.0%.

In other words, in the factor $e^{-\sigma t}$, $\sigma$ tells you the number of "$\frac{1}{e}$ steps" by which the amplitude is reduced per unit time. And the number of steps per unit time (steps per second in SI units) can reasonably be characterized as a type of frequency.

For a constant-envelope function such as sine or cosine functions, any amount of amplitude decay over time suffices to drive the function to zero after infinite time, so any value of $\sigma$ greater than zero in the function $e^{-\sigma t}$ will prevent the integral in the Laplace transform of a purely sinusoidal function from blowing up. But other time-domain functions, such as an exponentially increasing function, require multiplication by a more-sharply decaying factor (that is, a larger value of $\sigma$) to keep the integration result finite. For this reason, you'll often see a "region of convergence" (ROC) specified for the Laplace transform of a particular time-domain function $f(t)$.

The ROC is generally specified as a range of values of $s$ for which the Laplace transform integral converges, such as $s > 0$ for pure sine and cosine functions. But $s = \sigma + i\omega$ is complex, so what exactly does $s > 0$ mean? Since it is the real part of $s$ that determines the rate of decay of the factor $e^{-\sigma t}$, you can interpret $s > 0$ to mean $\sigma > 0$. Hence the results of the integration are finite as long as the real part $\sigma$ of the complex frequency $s$ exceeds the specified value.

You may be wondering whether time-domain functions exist for which no value of $\sigma$ is sufficient to cause the integral in the Laplace transform to converge. Yes, they do, and the time-domain function $f(t) = e^{t^2}$ is an example of such a function for which the Laplace transform does not exist. But the Laplace transform definitely does exist for any function that is of "exponential order" and that is "piecewise continuous".[6]

And exactly what does it mean for a function $f(t)$ to be of exponential order? Just that there exists an exponential function $Me^{c_0 t}$ for some constants $M$ and $c_0$ that is larger than the absolute value of $f(t)$ after some time $t = T$. That's because if such a function exists, the exponential term $e^{-st}$ in the Laplace transform ensures that the product $f(t)e^{-st}$ does not grow without limit for some value of $s$.

The other property, piecewise continuity, defines a function as piecewise continuous over an interval as long as the function is continuous (that is, has no discontinuities) over a finite number of subintervals, and any discontinuities at the ends of the subintervals are finite (that is, the difference between the

---

[6] The Laplace transform of a function may exist even if these two conditions are not satisfied, but if they are satisfied, then the Laplace transform is guaranteed to exist. In other words, these two conditions are sufficient but not necessary for the Laplace transform to exist.

function's value just before and just after each discontinuity cannot be infinitely large). This means that if $f(t)$ is piecewise continuous the contributions to the Laplace integral of discontinuities between subintervals must be finite. So as long as $f(t)$ is piecewise continuous and also of exponential order, the integral in the Laplace transform will converge.

Happily, most functions of interest in practical applications of the Laplace transform meet both of these conditions, which means that the Laplace transform of such functions does exist, because the integral of $f(t)e^{-st}$ does converge. In that case, the relationship between the Laplace transform and the Fourier transform can be summarized like this: if they both exist, the two-sided Laplace transform of the time-domain function $f(t)$ produces the same result as the Fourier transform of the modified function $f(t)e^{-\sigma t}$.

And how does the one-sided Laplace transform, with its lower integration limit of $0^-$, relate to the Fourier transform? For functions that are continuous across time $t = 0$ (that is, functions that do not have a discontinuity at $t = 0$), the one-sided Laplace transform of a time-domain function $f(t)$ produces the same result as the Fourier transform (if it exists) of the modified function $f(t)u(t)e^{-\sigma t}$, in which $u(t)$ represents the Heaviside unit-step function. The Heaviside function has amplitude zero for all negative time $t < 0$ and amplitude one for $t \geq 0$, so multiplying the time function $f(t)$ by this step function $u(t)$ has the effect of setting $f(t) = 0$ for all negative time.

To see that the one-sided Laplace transform of $f(t)$ and the Fourier transform of $f(t)u(t)e^{-\sigma t}$ produce the same result, consider the case of $f(t) = \cos(\omega_1 t)$. The Fourier transform of $f(t)u(t)e^{-\sigma t}$ looks like this:

$$F(\omega) = \int_{-\infty}^{+\infty} f(t)u(t)e^{-\sigma t}e^{-i\omega t}\,dt,$$

which for $f(t) = \cos(\omega_1 t)$ is

$$F(\omega) = \int_{-\infty}^{+\infty} (\cos(\omega_1 t))u(t)e^{-\sigma t}e^{-i\omega t}\,dt$$

$$= \int_{-\infty}^{+\infty} \left(\frac{e^{i\omega_1 t} + e^{-i\omega_1 t}}{2}\right)u(t)e^{-\sigma t}e^{-i\omega t}\,dt$$

$$= \int_{0}^{+\infty} \left(\frac{e^{i\omega_1 t} + e^{-i\omega_1 t}}{2}\right)e^{-\sigma t}e^{-i\omega t}\,dt$$

$$= \int_{0}^{+\infty} \left(\frac{e^{i(\omega_1 - \omega)t} + e^{-i(\omega_1 + \omega)t}}{2}\right)e^{-\sigma t}\,dt$$

$$= \int_0^{+\infty} \frac{e^{-[\sigma + i(\omega - \omega_1)]t} + e^{-[\sigma + i(\omega_1 + \omega)]t}}{2} dt$$

$$= -\frac{1}{2} \left[ \frac{e^{-[\sigma + i(\omega - \omega_1)]t}}{\sigma + i(\omega - \omega_1)} + \frac{e^{-[\sigma + i(\omega + \omega_1)]t}}{\sigma + i(\omega + \omega_1)} \right] \Bigg|_0^{+\infty}.$$

Inserting the upper limit of $+\infty$ for $t$ into the exponentials makes the numerator for both terms equal to zero as long as $\sigma > 0$, and inserting the lower limit of $0$ for $t$ makes both numerators equal to one (with a minus sign since this is the lower limit). So this evaluates to

$$F(\omega) = \frac{1}{2} \left[ \frac{1}{\sigma + i(\omega - \omega_1)} + \frac{1}{\sigma + i(\omega + \omega_1)} \right]$$

$$= \frac{1}{2} \left[ \frac{1}{(\sigma + i\omega) - i\omega_1} + \frac{1}{(\sigma + i\omega) + i\omega_1} \right]$$

$$= \frac{1}{2} \left[ \frac{(\sigma + i\omega) + i\omega_1}{[(\sigma + i\omega) - i\omega_1][(\sigma + i\omega) + i\omega_1]} \right.$$

$$\left. + \frac{(\sigma + i\omega) - i\omega_1}{[(\sigma + i\omega) + i\omega_1][(\sigma + i\omega) - i\omega_1]} \right]$$

$$= \frac{1}{2} \left[ \frac{(\sigma + i\omega) + i\omega_1 + (\sigma + i\omega) - i\omega_1}{(\sigma + i\omega)^2 + \omega_1^2} \right] = \frac{1}{2} \left[ \frac{2(\sigma + i\omega)}{(\sigma + i\omega)^2 + \omega_1^2} \right].$$

Thus

$$F(\omega) = \frac{\sigma + i\omega}{(\sigma + i\omega)^2 + \omega_1^2} \tag{1.22}$$

as long as $\sigma > 0$.

In Chapter 2, you can find the results $F(s)$ of the unilateral Laplace transform for a variety of time-domain functions, including the pure cosine function $f(t) = \cos(\omega_1 t)$. As you'll see, in that case $F(s)$ is given by

$$F(s) = \frac{s}{s^2 + \omega_1^2} \tag{1.23}$$

with a ROC of $s > 0$, which agrees with the Fourier transform results given by Eq. 1.22 since $s = \sigma + i\omega$.

As this example demonstrates, the unilateral Laplace transform finds the sine and cosine functions that mix together to give $f(t)e^{-\sigma t}$ for positive time $(t > 0)$. But the Laplace transform is "invertible," which means that the original time-domain function $f(t)$ can be recovered from $F(s)$ using an inverse transform. So if the time-domain function $f(t)$ is modified by the factor $e^{-\sigma t}$ when $F(s)$ is found using the forward Laplace transform, how does the inverse Laplace transform recover the original time-domain function $f(t)$ rather than $f(t)e^{-\sigma t}$? As you will see in Section 1.6, the inverse Laplace transform includes a factor of $e^{+\sigma t}$, and that factor compensates for the $e^{-\sigma t}$ factor by which $f(t)$ was modified in the forward-transform process.

One useful characteristic of $F(s)$, the result of taking the Laplace transform of a piecewise, exponential-order time-domain function, is that $F(s)$ must approach zero as $s$ approaches infinity:

$$\lim_{s \to \infty} F(s) = 0. \tag{1.24}$$

So whenever you take the Laplace transform of a time-domain function, it's a good idea to check that your expression for $F(s)$ goes to zero in the limit as $s$ goes to infinity.

An alternative perspective of the Laplace transform that you may find helpful considers the real exponential factor $e^{-\sigma t}$ as a modifier of the sinusoidal basis functions ($e^{i\omega t}$) rather than the time-domain function $f(t)$. In that case, the basis functions are not the constant-amplitude sine and cosine functions illustrated in Fig. 1.3 and described in Section 1.4; instead they are complex sinusoids weighted by the factor $e^{-\sigma t}$, an example of which is shown in Fig. 1.14.

If you take this perspective, you should note that these exponentially growing or decaying basis functions have a very important difference from purely sinusoidal basis functions: they are not orthogonal to one another. That is, unlike sine and cosine functions, or cosine functions at different frequencies, the "direction" of one of these basis functions in the abstract vector space they inhabit is not perpendicular to the direction of the other basis functions (the inner product between them is not always zero). That means that the results of the inner-product operation between each basis function $e^{(\sigma + i\omega)t}$ and $f(t)$ cannot be interpreted simply as the "amount" of that basis function (and no other) in $f(t)$, as is done for the vector components in Eq. 1.19 using orthogonal basis vectors. What does this mean about the output $F(s)$ of the Laplace transform? Only that the process of reconstructing the time-domain function $f(t)$ from $F(s)$, specifically the integration over the complex frequency parameter $s$, must be performed in a certain way. That process is described in detail in Section 1.6.

Students often find it helpful to see plots of the results of Laplace transforms, but since the Laplace parameter $s$ is complex, showing $F(s)$ as a function of both the real part $\sigma$ and the imaginary part $\omega$ of $s$ means that you're going to need another dimension. A third dimension can be shown by "laying

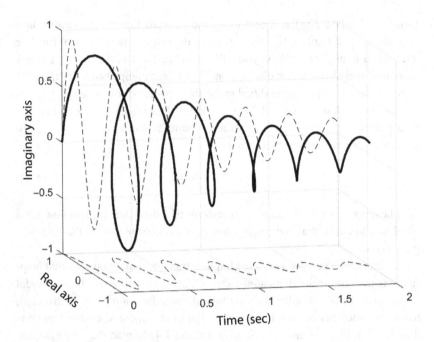

Figure 1.14 A complex sinusoidal basis vector modified by a real exponential factor $e^{-\sigma t}$ with $\omega = 20$ rad/sec and $\sigma = 1$/sec.

the $s$-plane on the floor" and adding another axis perpendicular to that plane, as in Fig. 1.15. In this figure, the magnitude of the complex frequency function $F(s)$ is plotted on the new (vertical) axis; the $F(s)$ used in this figure is the unilateral Laplace transform of the pure cosine function $f(t) = \cos(\omega_1 t)$.

If you look closely at Fig. 1.15, you can see that $F(s)$ differs from the frequency spectrum of a pure cosine function found by the Fourier transform, which consists of Dirac delta functions at angular frequencies $-\omega_1$ and $+\omega_1$. The peaks in $F(s)$ at $-\omega_1$ and $+\omega_1$ are clearly not infinitely tall, nor are they infinitely thin. There are two reasons for these differences between $F(s)$ and $F(\omega)$: in the Laplace transform, the time-domain function $f(t)$ is modified by the real exponential factor $e^{-\sigma t}$, and the integral in the unilateral Laplace transform has a lower limit of $t = 0^-$ rather than $t = -\infty$. You can think of this lower-limit difference as arising either from multiplying the time-domain function $f(t)$ by the Heaviside step function $u(t)$ prior to integration or from using the unilateral rather than the bilateral Laplace transform. Either way, the result is the same: additional frequency components are needed to ensure that $f(t)$ has zero amplitude for all negative time ($t < 0$) and to produce the

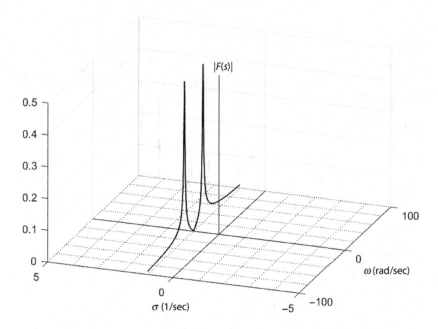

Figure 1.15 Three-dimensional $s$-plane plot of the magnitude of the Laplace transform result $|F(s)|$ for $f(t) = \cos(\omega_1 t)$ with $\omega_1 = 20\,\text{rad/sec}$ and $\sigma = 1/\text{sec}$.

decreasing amplitude of the $e^{-\sigma t}$ factor. At larger values of $\sigma$, the decrease in amplitude of the modified time-domain function is faster, and narrowing a time-domain function increases its width in the frequency domain.

You can see this happening in Fig. 1.16, which shows $F(s)$ for the same time-domain function $f(t)$ as the previous figure, but with increasing values of $\sigma$. At larger values of $\sigma$, the frequency peaks are both wider and not as tall due to the faster decay rates produced by larger $\sigma$. The key point for plots such as these is that within the ROC of the Laplace transform, the Fourier transform of the modified time-domain function $f(t)u(t)e^{-\sigma t}$ and the unilateral Laplace transform of $f(t)$ yield identical results.

Another type of plot you may also encounter in books and websites dealing with the Laplace transform shows $F(s)$ as a continuous surface above the $s$-plane, as in Fig. 1.17. In this plot, the magnitude of $F(s)$, the Laplace transform of the time-domain function $f(t) = \cos(\omega_1 t)$, is shown for values of $\sigma$ ranging from 0.5/sec to 5/sec and values of $\omega$ ranging from $-50$ rad/sec to 50 rad/sec.

In some texts, you may also see $s$-plane plots of $F(s)$ that are not restricted to the ROC of the integral in the Laplace transform. This causes some readers

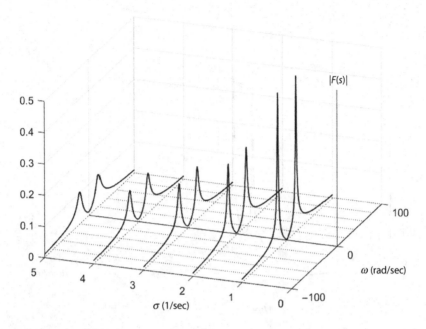

Figure 1.16 $|F(s)|$ for $f(t) = \cos(\omega_1 t)$ with $\omega_1 = 20$ rad/sec and $\sigma = 1$/sec, 2/sec, 3/sec, 4/sec, and 5/sec.

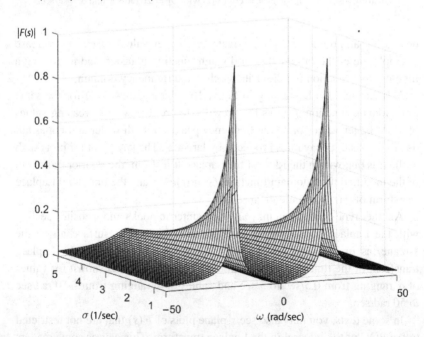

Figure 1.17 Surface plot of $|F(s)|$ for $f(t) = \cos(\omega_1 t)$ with $\omega = 20$ rad/sec and $\sigma$ increasing in small steps from 0.5/sec to 5/sec.

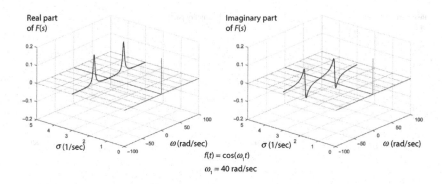

Figure 1.18 Real and imaginary parts of $F(s)$ for $f(t) = \cos(\omega_1 t)$.

to wonder how the behavior of $F(s)$ can be determined in regions in which the defining integral does not converge. The answer is that a power-series process called "analytic continuation" can be used to extend the domain of a complex function to additional points; in this case those additional points are values of $s$ outside the ROC. You can find links to helpful resources with more detail about analytic continuation on this book's website.

When you see three-dimensional plots of the results $F(s)$ of the Laplace transform, it's important to remember that both the generalized frequency $s$ and the Laplace transform result $F(s)$ may be complex. So although you're seeing the real part ($\sigma$) and imaginary part ($\omega$) of $s$ on the $x$- and $y$-axes, the $z$-axis can't show both the real part and the imaginary part of $F(s)$. That's why the magnitude of $F(s)$ is plotted on the $z$-axis in Figs. 1.15, 1.16, and 1.17.

As an alternative to plotting the magnitude of $F(s)$ on the $z$-axis, you may also see three-dimensional $s$-plane plots with the real or imaginary part of $F(s)$ on the $z$-axis, or perhaps the phase of $F(s)$, given by $\tan^{-1}\{\text{Im}[F(s)]/\text{Re}[F(s)]\}$. It's also common to show a single slice through the real or imaginary part of $F(s)$ taken along a line of constant $\sigma$, as in Fig. 1.18. In this case, slices through the real and imaginary parts of the Laplace transform of $f(t) = \cos(\omega_1 t)$ with $\omega_1 = 40$ rad/sec have been taken along the $\sigma = 3/\text{sec}$ line.

For comparison, slices through the real and imaginary parts of the Laplace transform $F(s)$ along the $\sigma = 3/\text{sec}$ line for $f(t) = \sin(\omega_1 t)$ are shown in Fig. 1.19. The important point is that when you encounter such plots you should keep in mind that the quantity being plotted on the vertical axis is complex, so it's typically the magnitude, real part, imaginary part, or phase of $F(s)$ (but not all four) that appears on the plot.

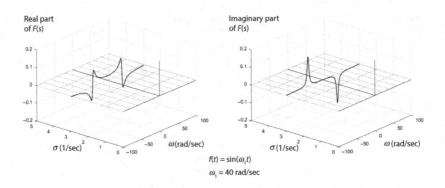

Figure 1.19 Real and imaginary parts of $F(s)$ for $f(t) = \sin(\omega_1 t)$.

Another $s$-plane plot that's often shown in engineering applications of the Laplace transform is the "pole-zero" plot. In these plots, values of $s$ at which $F(s)$ approaches $\infty$ (called poles) are marked with an "x," and values of $s$ at which $F(s)$ equals zero (called zeros) are marked with an "o." It is straightforward to find these locations (that is, values of $\sigma$ and $\omega$) if you express $F(s)$ as a ratio of polynomials, because the roots of the numerator polynomial are the zeros and the roots of the denominator polynomial are the poles.

Here is an example: as shown in Eq. 1.23 and derived in Chapter 2, the Laplace transform of the cosine function $f(t) = \cos(\omega_1 t)$ is $F(s) = \frac{s}{s^2 + \omega_1^2}$. That means that there's a single zero at $s = 0$ (the origin of the $s$-plane), because the numerator of $F(s)$ is zero when $s = 0$. As for poles, there are two, because the denominator of $F(s)$ goes to zero when $s = i\omega_1$ and when $s = -i\omega_1$. So the poles in this case both lie along the $\sigma = 0$ line, one at $\omega = \omega_1$ and the other at $\omega = -\omega_1$. You can see the pole-zero diagram for this case in Fig. 1.20a.

The number and location of the $s$-plane poles and zeros depends on the function $f(t)$ and its Laplace transform $F(s)$. For example, as you'll see in Chapter 2, the Laplace transform of the function $f(t) = e^{at}$ is $F(s) = \frac{1}{s-a}$, which has no zeros (since the numerator of $F(s)$ never equals zero) and a single, real pole at $s = a$ (which makes the denominator zero). A pole-zero diagram for this case is shown in Fig. 1.20b.

Pole-zero diagrams are especially useful in assessing and specifying the performance of signal-processing and control systems; you can see additional examples in one of the chapter-end problems and its online solution. And there are some very useful rules relating the ROC to the poles of the Laplace

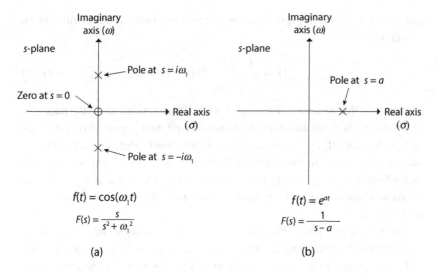

Figure 1.20 Pole-zero plots for (a) $F(s) = \frac{s}{s^2+\omega_1^2}$ and (b) $F(s) = \frac{1}{s-a}$.

transform, as described in the "Regions of Convergence" document on this book's website.

## 1.6 Inverse Transforms

It may help you to develop a deeper understanding of the meaning of the frequency-domain function $F(\omega)$ produced by the Fourier transform and the complex-frequency function $F(s)$ produced by the Laplace transform by considering the inverse transform. As mentioned in Section 1.3, the inverse Fourier transform "goes the other way" by converting the frequency spectrum $F(\omega)$ into the time-domain function $f(t)$. Likewise, the inverse Laplace transform converts the complex-frequency function $F(s)$ into the time-domain function $f(t)$; you can see how that process works for each of the Laplace transform examples in the next chapter.

To understand the differences between the forward and inverse Fourier and Laplace transforms, recall that the mathematical expression for the forward Fourier transform can be written like this:

$$F(\omega) = \int_{-\infty}^{+\infty} f(t)e^{-i\omega t}\,dt \qquad (1.7)$$

and the mathematical expression for the inverse Fourier transform can be written like this:

$$f(t) = \frac{1}{2\pi} \int_{-\infty}^{\infty} F(\omega)e^{i\omega t}\,d\omega. \qquad (1.16)$$

It's apparent that these transforms go in opposite directions, since the positions of the functions $F(\omega)$ and $f(t)$ are switched, and the integration is performed in the domain of the function being transformed. But, as mentioned in Section 1.3, there are two other differences: the sign of exponent in the multiplicative term $e^{i\omega t}$ is positive in the inverse transform and negative in the forward transform, and the inverse transform has a factor of $\frac{1}{2\pi}$ in front of the integral.

You should understand that the exact form of these equations is a matter of convention rather than fundamental mathematics, so you may encounter books and websites in which the sign of the complex exponential is negative in the inverse Fourier transform and positive in the forward transform; you may also see the factor of $\frac{1}{2\pi}$ applied to the forward rather than the inverse transform (or shared between them, as described below).

The role of the sign of the exponent in the term $e^{i\omega t}$ in the inverse Fourier transform can be understood by remembering that the inverse transform synthesizes the time-domain function $f(t)$ by weighting and combining sinusoidal basis functions such as $\cos(\omega t)$ and $\sin(\omega t)$. Recall also that Euler's relation tells you that the sign of the exponent in the expression $e^{\pm i\omega t}$ tells you the sign of the sine-function basis components:

$$e^{\pm i\omega t} = \cos(\omega t) \pm i\sin(\omega t), \qquad (1.4)$$

so a positive sign in the exponent means that the sine component is positive. In other words, the contribution of the sine-function components to $f(t)$ is zero at the starting time ($t = 0$) and is positive for the first half-cycle in positive time ($t > 0$).

As mentioned above, it is possible to define the inverse Fourier transform as having a negative sign in the exponential term – in that case, the contributions to the time-domain function $f(t)$ from the sine-function components are negative for the first half-cycle after $t = 0$. Using that convention, the forward Fourier transform must have a positive sign in the exponent of $e^{i\omega t}$ to ensure that the inverse transform of the function $F(\omega)$ produced by the forward transform returns the original time-domain function $f(t)$.

To see how that works, and to understand why there is a factor of $\frac{1}{2\pi}$ in front of the integral in the inverse Fourier transform, write the constant "$c_1$" rather

than $\frac{1}{2\pi}$ in the inverse transform. That makes the expression for the inverse
Fourier transform of $f(t)$ look like this:

$$f(t) = c_1 \int_{-\infty}^{\infty} F(\omega)e^{i\omega t} d\omega.$$

Now observe what happens when you take the inverse Fourier transform of the
forward Fourier-transform output $F(\omega)$:

$$f(t) = c_1 \int_{\omega=-\infty}^{\infty} F(\omega)e^{i\omega t} d\omega = c_1 \int_{\omega=-\infty}^{\infty} \left[\int_{\tau=-\infty}^{+\infty} f(\tau)e^{-i\omega\tau} d\tau\right] e^{i\omega t} d\omega,$$

in which the time variable $\tau$ has been used as the argument of the function
$f(\tau)$ to differentiate it from the time variable $t$ of the inverse transform.

Since the exponential term $e^{-i\omega t}$ does not depend on the separate time
variable $\tau$, it can be moved inside the integral over $\tau$, giving

$$f(t) = c_1 \int_{\omega=-\infty}^{\infty} \int_{\tau=-\infty}^{+\infty} f(\tau)e^{-i\omega\tau}e^{i\omega t} d\tau d\omega$$

$$= c_1 \int_{\omega=-\infty}^{\infty} \int_{\tau=-\infty}^{+\infty} f(\tau)e^{-i\omega(\tau-t)} d\tau d\omega.$$

Switching the order of integration (which you can do as long as the integrals
converge) makes this

$$f(t) = c_1 \int_{\tau=-\infty}^{\infty} \int_{\omega=-\infty}^{+\infty} f(\tau)e^{-i\omega(\tau-t)} d\omega d\tau$$

$$= c_1 \int_{\tau=-\infty}^{\infty} f(\tau) \left[\int_{\omega=-\infty}^{+\infty} e^{-i\omega(\tau-t)} d\omega\right] d\tau.$$

As described in Section 1.3, the integral inside the square brackets can be
replaced by a Dirac delta function

$$\int_{\omega=-\infty}^{+\infty} e^{-i\omega(\tau-t)} d\omega = 2\pi\delta(\tau - t),$$

so

$$f(t) = c_1 \int_{\tau=-\infty}^{+\infty} f(\tau) [2\pi\delta(\tau - t)] d\tau$$

$$= c_1(2\pi) \int_{\tau=-\infty}^{+\infty} f(\tau)\delta(\tau - t) d\tau = c_1(2\pi)f(t),$$

in which the sifting property of the delta function has been used. Dividing both sides by $f(t)$ and solving for the constant $c_1$ gives

$$1 = c_1(2\pi).$$

$$c_1 = \frac{1}{2\pi}.$$

So if you use the form of the forward Fourier transform shown in Eq. 1.7, the inverse Fourier transform must include the multiplicative constant $\frac{1}{2\pi}$ if you want the inverse transform of $F(\omega)$ to return the original time-domain function $f(t)$. You should note that many authors choose to "split the constant" between the forward and inverse transform by defining the forward Fourier transform as

$$F(\omega) = \frac{1}{\sqrt{2\pi}} \int_{-\infty}^{+\infty} f(t)e^{-i\omega t}\,dt$$

and the inverse Fourier transform as

$$f(t) = \frac{1}{\sqrt{2\pi}} \int_{-\infty}^{\infty} F(\omega)e^{i\omega t}\,d\omega$$

while others attach the factor of $\frac{1}{2\pi}$ to the forward rather than the inverse transform. You may use whichever of these approaches you prefer, as long as the product of the constants you include in the forward and inverse transform is $\frac{1}{2\pi}$.

You may also encounter texts which employ the linear frequency $f$ (sometimes written as $v$) rather than the angular frequency $\omega$ in defining the Fourier transform and its inverse. In that case, the forward Fourier transform can be written as

$$F(v) = \int_{-\infty}^{+\infty} f(t)e^{-i2\pi v t}\,dt$$

and the inverse Fourier transform as

$$f(t) = \int_{-\infty}^{\infty} F(v)e^{i2\pi v t}\,dv.$$

No matter which form of the Fourier transform you use, having a good understanding of the relationship between the forward and inverse Fourier transforms can definitely make it easier to comprehend the form and function of the inverse Laplace transform and its relationship to the forward Laplace transform. But since the Laplace transform involves the complex frequency $s$ rather than the real frequency $\omega$, there are some important differences between the inverse Laplace transform and the inverse Fourier transform.

Those differences are readily apparent in the mathematical expression for the inverse Laplace transform, often written like this:

$$f(t) = \mathcal{L}^{-1}[F(s)] = \frac{1}{2\pi i} \int_{\sigma-i\infty}^{\sigma+i\infty} F(s)e^{st}ds, \qquad (1.25)$$

in which the symbol $\mathcal{L}^{-1}$ represents the inverse Laplace transform operator and, as usual, $s = \sigma + i\omega$ represents the complex frequency. If the $s$-domain function $F(s)$ used in this equation is the result of the unilateral Laplace transform, the time-domain function $f(t)$ synthesized by this approach is zero for times $t < 0$. For times $t > 0$, this equation tells you how to find the time function $f(t)$ from the complex-frequency function $F(s)$, just as the inverse Fourier transform tells you how to find $f(t)$ from $F(\omega)$.

Comparing Eq. 1.25 for the inverse Laplace transform to Eq. 1.16 for the inverse Fourier transform, you may notice that beyond replacing the real frequency $\omega$ with the complex frequency $s$, there's a factor of $i$ in the denominator, and the limits of integration are complex, with real part $\sigma$ and imaginary parts of $\pm\infty$.

Those differences arise from the fact that the integration in the inverse Laplace transform operates in the complex-frequency domain. That's a form of integration that you may not have seen before – it involves a line integral in the complex plane, called contour integration. You can find a brief discussion of contour integration with references and links to helpful websites on this book's website, but you can understand Eq. 1.25 for the inverse Laplace transform by realizing that this integral operates on a line of constant $\sigma$ extending from $\omega = -\infty$ to $\omega = +\infty$. And since $s = \sigma + i\omega$, $\frac{ds}{d\omega} = i$, and $d\omega = \frac{ds}{i}$, which accounts for the factor of $i$ in the denominator of Eq. 1.25.

Does this equation for the inverse Laplace transform give the time-domain function $f(t)$ for a given value of $\sigma$? Yes it does, and you can see this process in action for each of the Laplace-transform examples in the next chapter.

Although it's worth your time to make sure you understand how the inverse Laplace transform is implemented by Eq. 1.25, in practice the inverse transform is often found by other methods. One common approach when you know the complex-frequency function $F(s)$ and you wish to determine the time function $f(t)$ is simply to consult a table of Laplace transforms; if the function $F(s)$ in which you're interested is included in the table, you can read off the corresponding function $f(t)$. Of course, tables of Laplace transforms typically contain only basic functions, but even if your $F(s)$ is not listed, it may be possible to use the Laplace-transform characteristics discussed in Chapter 3 to relate a complicated function to those basic functions. In some cases, techniques such as partial fractions and series methods may be helpful in converting a complicated function $F(s)$ into a group of simpler functions, and those simpler functions may appear in the table. This book's website has an overview of the partial-fraction technique along with helpful references and links to relevant resources.

Before you move on to the examples of the Laplace transform presented in the next chapter, I strongly recommend that you work through the problems provided in the final section of this chapter. As you do so, remember that full interactive solutions for every problem are available on this book's website.

## 1.7 Problems

1. Make phasor diagrams to show how the counter-rotating phasors $e^{i\omega t}$ and $e^{-i\omega t}$ can be combined to produce the functions $\cos(\omega t)$ and $\sin(\omega t)$ as given by Eqs. 1.5 and 1.6.

2. Use the definition of the Fourier transform (Eq. 1.7) and the sifting property (Eq. 1.13) of the Dirac delta function $\delta(t)$ to find the frequency spectrum $F(\omega)$ of $\delta(t)$.

3. Find the frequency spectrum $F(\omega)$ of the constant time-domain function $f(t) = c$. Then find and sketch $F(\omega)$ for the time-limited function $f(t) = c$ between $t = -t_0$ and $t = +t_0$ and zero elsewhere.

4. Use the definition of the inverse Fourier transform (Eq. 1.16) to show that $f(t) = \cos(\omega t)$ is the inverse Fourier transform of $F(\omega)$ given by Eq. 1.12 and that $f(t) = \sin(\omega t)$ is the inverse Fourier transform of $F(\omega)$ given by Eq. 1.15.

5. If vector $\vec{A} = 3\hat{\imath} - 2\hat{\jmath} + \hat{k}$ and vector $\vec{B} = 6\hat{\jmath} - 3\hat{k}$, what are the magnitudes $|\vec{A}|$ and $|\vec{B}|$, and what is the value of the scalar product $\vec{A} \circ \vec{B}$?

6. For the vectors $\vec{A}$ and $\vec{B}$ defined in the previous problem, use Eq. 1.18 and the results of the previous problem to find the angle between $\vec{A}$ and $\vec{B}$.

7. The Legendre functions of the first kind, also called Legendre polynomials, are a set of orthogonal functions that find application in a variety of physics and engineering problems. The first four of these functions are

$$P_0(x) = 1 \qquad P_2(x) = \frac{1}{2}(3x^2 - 1)$$

$$P_1(x) = x \qquad P_3(x) = \frac{1}{2}(5x^3 - 3x).$$

Show that these four functions are orthogonal to one another over the interval from $x = -1$ to $x = +1$.

8. Find the Fourier transform $F(\omega)$ of the modified function $f(t)u(t)e^{-\sigma t}$ for $f(t) = \sin(\omega_1 t)$ following the approach used in Section 1.5 for the modified cosine function. Compare your result to Eq. 2.17 in Chapter 2 for the unilateral Laplace transform $F(s)$ of a sine function.

9. Show that the limit as $s$ approaches infinity for $F(s) = \frac{s}{s^2 + \omega_1^2}$ (Eq. 1.23) is zero, in accordance with Eq. 1.24.

10. Make pole-zero diagrams for the $s$-domain functions $F(s) = \frac{3s + 2}{s^2 - s - 2}$ and $F(s) = \frac{2s}{s^2 + 4s + 13}$.

# 2

# Laplace-Transform Examples

Becoming familiar with the Laplace transform $F(s)$ of basic time-domain functions $f(t)$ such as exponentials, sinusoids, powers of $t$, and hyperbolic functions can be immensely useful in a variety of applications. That is because many of the more complicated functions that describe the behavior of real-world systems and that appear in differential equations can be synthesized as a mixture of these basic functions. And although there are dozens of books and websites that show you how to find the Laplace transform of such functions, much harder to find are explanations that help you achieve an intuitive understanding of why $F(s)$ takes the form it does, that is, an understanding that goes beyond "That's what the integral gives". So the goal of this chapter is not just to show you the Laplace transforms of some basic functions, but to provide explanations that will help you see why those transforms make sense.

Another reason for using simple time-domain functions as the examples in this chapter is that simple functions illustrate the process of taking the Laplace transform with minimal mathematical distraction and can help you understand the meaning of the result $F(s)$. But even these simple time-domain functions may be combined to produce more complicated functions, and the characteristics of the Laplace transform described in Chapter 3 often allow you to determine $F(s)$ for those functions by combining the transform results of simpler functions. There is no simpler time-domain function than a real, constant value that exists over all time, so that's the first example (Section 2.1), followed by exponential functions in Section 2.2, sinusoidal functions in Section 2.3, power functions of the form $t^n$ in Section 2.4, and hyperbolic functions in Section 2.5. As in every chapter, the final section contains a set of problems that allow you to check your understanding of the material presented in this chapter.

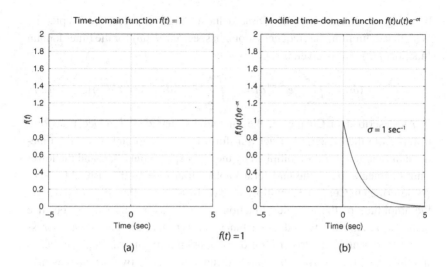

Figure 2.1 (a) Constant time-domain function $f(t)$ and (b) modified $f(t)$.

## 2.1 Constant Functions

An all-time constant function can be written as

$$f(t) = c, \tag{2.1}$$

in which $c$ represents the unchanging, real value of the function from $t = -\infty$ to $t = +\infty$.

This time-domain function can be considered to have a frequency of zero, since frequency is determined by the number of cycles per second (or radians per second for angular frequency), and there are no cycles in this case. Electrical engineers often call an unchanging signal described by $f(t) = c$ a "DC" signal to distinguish it from an "AC" signal that varies from positive to negative over time.[1] A plot of $f(t)$ for $c = 1$ over a 10-second time window is shown in Fig. 2.1a.

Since this function is comprised of a single frequency, its Fourier transform is a single, infinitely narrow spike at $\omega = 0$. But that spike is infinitely tall, because the integral in the Fourier transform of $f(t) = c$ does not converge – the area under the curve of $f(t)$ multiplied by $e^{-i\omega t}$ with $\omega = 0$ over all time is infinite. Hence the Fourier transform of this function can be found only

---

[1] "DC" was originally coined as an abbreviation for "direct current" in which current flows steadily in one direction, while AC was shorthand for "alternating current" in which current reverses direction periodically. You may see these terms applied to quantities other than current.

by using the definition of the Dirac delta function (Eq. 1.10 in Chapter 1). With that definition, the frequency-domain spectrum $F(\omega)$ for the time-domain function $f(t) = c$ is given by $c[2\pi\delta(\omega)]$:

$$F(\omega) = \int_{-\infty}^{\infty} ce^{-i\omega t}dt = c\int_{-\infty}^{\infty} e^{-i\omega t}dt = c[2\pi\delta(\omega)]. \qquad (2.2)$$

As described in Chapter 1, the unilateral (one-sided) Laplace transform differs from the bilateral Fourier transform in two important ways: the time-domain function $f(t)$ is multiplied by the real exponential $e^{-\sigma t}$, and the lower limit of integration in the unilateral Laplace transform is $0^-$ instead of $-\infty$. As described in Chapter 1, multiplying $f(t)$ by $e^{-\sigma t}$ has the effect of reducing the amplitude of the modified function $f(t)e^{-\sigma t}$ over time for any positive value of $\sigma$. And for time-domain functions $f(t)$ that are continuous across time $t = 0$, setting the lower limit of integration to $0^-$ has the same effect as multiplying $f(t)$ by $u(t)$, the unit-step function. Those two differences may allow the one-sided Laplace transform to converge when the Fourier transform does not. But as described in Section 1.5, that convergence may occur only for a certain range of values of $\sigma$; bear in mind that the region of convergence (ROC) depends on the behavior of $f(t)$ over time. A plot of the modified function $f(t)u(t)e^{-\sigma t}$ for $f(t) = 1$ and $\sigma = 1/\mathrm{sec}$ is shown in Fig. 2.1b.

For $f(t) = c$, the unilateral Laplace transform looks like this:

$$F(s) = \int_{0^-}^{+\infty} f(t)e^{-st}dt = \int_{0}^{+\infty} ce^{-st}dt, \qquad (2.3)$$

in which the lower limit of integration can be written as 0 rather than $0^-$ in the second integral, since in this case the time-domain function $f(t)$ is continuous across time $t = 0$.

The improper integral in Eq. 2.3 may be made to converge, as you can see by calling the upper limit of integration $\tau$ and taking the limit as $\tau$ goes to infinity:

$$F(s) = \int_{0}^{+\infty} ce^{-st}dt = \lim_{\tau\to\infty}\int_{0}^{\tau} ce^{-st}dt$$

$$= \lim_{\tau\to\infty}\left[c\frac{1}{-s}e^{-st}\right]\Big|_{0}^{\tau} = \lim_{\tau\to\infty}\left[\frac{c}{-s}e^{-s\tau} - \frac{c}{-s}e^{-s(0)}\right].$$

The first term in the final expression is zero as long as $s$ (specifically, its real part $\sigma$) is greater than zero. That leaves only the second term, so

$$F(s) = -\frac{c}{-s}e^{0} = \frac{c}{s} \qquad\qquad \mathrm{ROC}:\ s > 0. \qquad (2.4)$$

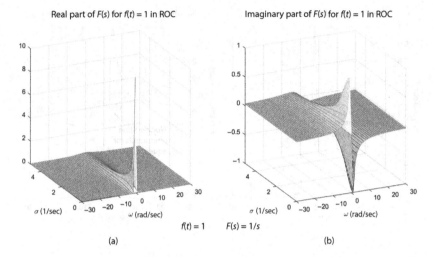

Figure 2.2 (a) Real and (b) imaginary parts of $F(s)$ for constant $f(t)$ in region of convergence.

So for $f(t) = c$, the integral of $f(t)e^{-st}$ over the range $t = 0$ to $t = \infty$ converges, the unilateral Laplace transform exists, and $F(s) = c/s$ as long as the real part of $s$ is greater than zero. In other words, even the slowest exponential decrease applied to a constant time-domain function is sufficient to prevent the area under the curve of $f(t)e^{-st}$ from becoming infinite over the time range of $t = 0$ to $t = \infty$.

At this point, it is customary to move on to other time-domain functions, but I think it's important to make sure you understand what Eq. 2.4 is telling you, and why the function $F(s) = c/s$ makes sense as the one-sided Laplace transform of a constant time-domain function.

To see that, start by looking at the three-dimensional $s$-plane plots of the real and imaginary parts of $F(s) = c/s$ in Fig. 2.2. For these three-dimensional plots, the constant $c$ is taken as unity, the range of $\sigma$ values begins just inside the region of convergence (at $\sigma = 0.1/\text{sec}$ in this case), and a range of $\omega$ values $(-30 \text{ rad/sec to} +30 \text{ rad/sec})$ is shown to give you an idea of the shape of the function $F(s) = c/s$.

As you can see in Fig. 2.2a, the real part of $F(s)$ is considerably larger than the imaginary part shown in Fig. 2.2b (note the difference in the vertical scales). To understand why that's true, remember that for any real time-domain function $f(t)$, the real part of the Laplace transform $F(s)$ tells you the amount of each cosine basis function $\cos(\omega t)$ present in the modified function $f(t)u(t)e^{-\sigma t}$, and the imaginary part of $F(s)$ tells you the amount of each sine

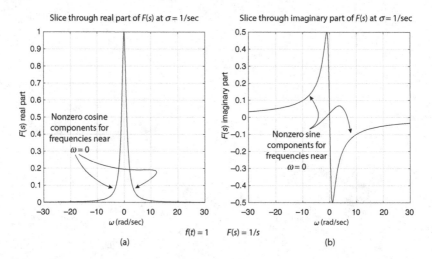

Figure 2.3 (a) Real and (b) imaginary parts of a slice through $F(s)$ at $\sigma = 1/\text{sec}$.

basis function $\sin(\omega t)$ in that same modified function. And in this case the unmodified constant function $f(t)$ can be considered to be a zero-frequency cosine function, since $\cos(\omega t) = 1$ for all time if $\omega = 0$.

But if the time-domain function was a pure cosine function before multiplication by the unit-step function $u(t)$ and the exponential decay function $e^{-\sigma t}$, why is the real part of $F(s)$ not a single spike at $\omega = 0$ and $\sigma = 0$? And why is the imaginary part of $F(s)$, which represents the amount of the $\sin(\omega t)$ function present in $f(t)$, not identically zero?

The answer to those questions is that the modified time-domain function shown in Fig. 2.1b has additional frequency components consisting of both cosine and sine functions. You can get a better idea of the amplitudes of those additional frequency components by looking at a two-dimensional slice at constant $\sigma$ through the three-dimensional $F(s)$ plot; one such slice is shown in Fig. 2.3, taken along the $\sigma = 1/\text{sec}$ line in the $s$-plane. The equations describing these slices can be found by separating out the real and imaginary parts of $F(s) = 1/s$:

$$F(s) = \frac{1}{s} = \frac{1}{\sigma + i\omega} = \frac{\sigma - i\omega}{\sigma^2 + \omega^2}$$

$$= \frac{\sigma}{\sigma^2 + \omega^2} + \frac{-i\omega}{\sigma^2 + \omega^2},$$

which means the real part of $F(s)$ is

$$\text{Re}[F(s)] = \frac{\sigma}{\sigma^2 + \omega^2} \tag{2.5}$$

and the imaginary part of $F(s)$ is

$$\text{Im}[F(s)] = \frac{-\omega}{\sigma^2 + \omega^2}. \tag{2.6}$$

Consider first the equation and the corresponding plot for the real part of $F(s)$. For a given value of $\sigma$, the amplitudes of the components on either side of $\omega = 0$ decrease rapidly with $\omega$ due to the $\omega^2$ term in the denominator, but those reduced-amplitude cosine components definitely have an effect on the time-domain function that results from adding together all the frequency components. Specifically, those additional frequency components cause the amplitude of the time-domain function to roll off over time in both directions (that is, as $t$ becomes more positive and more negative). And as you'll see when you read about the inverse Laplace transform of $F(s) = \frac{c}{s}$ later in this section, the shape of the real part of $F(s)$ given by Eq. 2.5 and shown in Fig. 2.3a produces the exact amount of roll-off in the time domain needed to match the modified function $f(t)e^{-\sigma t}$.

Of course, the $s$-domain function $F(s)$ represents the frequency components present not in $f(t)e^{-\sigma t}$, but in $f(t)u(t)e^{-\sigma t}$, which is zero for all negative time ($t < 0$). But all cosine basis functions are even, since $\cos(\omega t) = \cos(-\omega t)$, which means they are symmetric about $t = 0$. So you might surmise that it's not possible to synthesize an asymmetric function about $t = 0$, such as $f(t)u(t)e^{-\sigma t}$, using only cosine basis functions. That is correct, and that's where the imaginary part of $F(s)$ comes in.

As you can see in Eq. 2.6 and in Fig. 2.3b, the magnitudes of the frequency components given by the imaginary part of $F(s)$ also decrease in both the negative-$\omega$ and the positive-$\omega$ directions, albeit more slowly than in the real part of $F(s)$ since the imaginary part has an $\omega$ in the numerator. And since the imaginary part of $F(s)$ is an odd function ($\text{Im}[F(s)] = -\text{Im}[F(-s)]$), you might think that the positive-$\omega$ and negative-$\omega$ components will cancel one another. But remember that the imaginary part of $F(s)$ pertains to the sine basis functions, and sine functions are themselves odd, since $\sin(\omega t) = -\sin(-\omega t)$. So applying the component amplitudes given by the odd function $\text{Im}[F(s)]$ to the odd basis functions $\sin(\omega t)$ means that the frequency components on opposite sides of $\omega = 0$ have the same sign.

It might appear that those components will all be negative for positive time, since for $\omega < 0$ the amplitudes of $\text{Im}[F(s)]$ are positive and $\sin(\omega t)$ is negative (and vice-versa when $\omega > 0$). But the imaginary unit $i$ in front of the imaginary part of $F(s)$ multiplies the $i$ in front of the $\sin(\omega t)$ term of $e^{i\omega t}$ to produce an additional factor of $-1$. So the sine components for this $f(t)$ have the shape of $+\sin(\omega t)$, which is positive for the first half-cycle on the right (positive) side of $t = 0$ and negative for the first half-cycle on the left (negative) side of $t = 0$.

And what is the value of adding in these sine-function components to the cosine-function components contributed by the real part of $F(s)$? Just this: since the modified time-domain function $f(t)u(t)e^{-\sigma t}$ is zero for all negative time, these sine components serve to reduce the amplitude of the integrated result when combined with the cosine components for times $t < 0$. But for positive values of $t$, the sine components combine with the cosine components in just the right way to produce the constant value of the time-domain function $f(t)$. You can see exactly how this combination of sine and cosine components produces the desired function $f(t)$ in the discussion of the inverse Laplace transform later this this section.

If you're wondering about the maximum height of $F(s)$ in the three-dimensional $s$-plane plots shown in Fig. 2.2, those values are determined by the step size in $\sigma$ and $\omega$ used to produce the plots. That is because those step sizes determine how close a sample is taken to the infinitely high "pole" of $F(s)$ at $\sigma = 0$ and $\omega = 0$. In these plots, the sample spacing for $\sigma$ is 0.1/sec, and the sample spacing for $\omega$ is 1 rad/sec. The maximum value of the real part of $F(s)$ occurs at $\omega = 0$ and $\sigma = 0.1$/sec, so you can plug those values into Eq. 2.5, which gives

$$\mathrm{Re}[F(s)] = \frac{\sigma}{\sigma^2 + \omega^2} = \frac{0.1}{(0.1)^2 + (0)^2} = 10,$$

consistent with Fig. 2.2a.

For the imaginary part of $F(s)$, the value at $\omega = 0$ and $\sigma = 0.1$/sec is zero, and the maximum value occurs at $\omega = -1$ rad/sec and $\sigma = 0.1$/sec. Plugging those values into Eq. 2.6 gives

$$\mathrm{Im}[F(s)] = \frac{-\omega}{\sigma^2 + \omega^2} = \frac{-(-1)}{(0.1)^2 + (-1)^2} = 0.99,$$

consistent with Fig. 2.2b.

A similar analysis shows why the maximum value of $\mathrm{Re}[F(s)]$ is $+1$ and the maximum value of $\mathrm{Im}[F(s)]$ is 0.5 in the slice taken at $\sigma = 1$/sec shown in Fig. 2.3.

In the remainder of this section, you will see the inverse Laplace transform in action – that is, producing the time-domain function $f(t) = c$ (for $t > 0$) from the complex-frequency domain function $F(s) = c/s$. As explained in Section 1.6, that process involves integrating the product of the functions

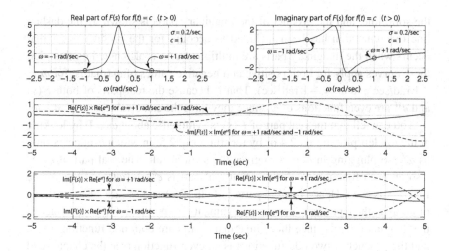

Figure 2.4 Real and imaginary parts of $F(s) = c/s$ and $F(s)e^{st}$.

$F(s)$ and $e^{st}$ over $\omega$ at a constant value of $\sigma$. But before seeing how that works, it is worth a bit of time and effort to make sure you understand the behavior of the individual frequency components of that product, and specifically the importance of both the real and the imaginary parts of the functions $F(s)$ and $e^{st}$.

To do that, take a look at the real and imaginary parts of $F(s) = c/s$ shown in the top portion of Fig. 2.4 for $\sigma = 0.2$/sec and $c = 1$. Students often express confusion over the role that the imaginary part of $F(s)$ plays in the synthesis of a purely real function such as $f(t) = c$, so you may find it helpful to examine the behavior of the product $F(s)e^{st}$ over time for a specific frequency.

You can see that behavior in the middle and bottom portions of Fig. 2.4 for the angular frequency $\omega = \pm 1$ rad/sec. Remember that both $F(s)$ and $e^{st}$ have real parts and imaginary parts, so the product of these two complex functions can also have real and imaginary parts. Specifically, multiplying the real part of $F(s)$ by the real part of $e^{st}$ produces a real result, as does the process of multiplying the imaginary part of $F(s)$ by the imaginary part of $e^{st}$ (since both of those imaginary parts include the imaginary unit "$i$", and the product $i \times i$ is the real number $-1$). But multiplying the real part of $F(s)$ by the imaginary part of $e^{st}$ gives an imaginary result, as does the process of multiplying the imaginary part of $F(s)$ by the real part of $e^{st}$ (since in those cases only one of the two terms includes the imaginary unit).

The real part of the product $F(s)e^{st}$ is shown in the middle portion of Fig. 2.4. As noted on the graph, the product of the real parts is shown by the solid line, and the product of the imaginary parts is shown by the dashed line (a minus sign has been included to account for the product $i \times i$ that occurs when the imaginary parts are multiplied). As is also noted on this plot, these curves apply for both positive and negative values of $\omega$ (that is, for $\omega = +1$ rad/sec and $\omega = -1$ rad/sec). That is because the real parts of both $F(s)$ and $e^{st}$ are even functions of $\omega$, since they have the same value at $\omega$ as at $-\omega$.

You can tell that the real part of $F(s)$ is symmetric about $\omega = 0$ by looking at the top left plot in Fig. 2.4 or by recalling Eq. 2.5, in which $\omega$ appears only as $\omega^2$, so plugging in $\omega$ or $-\omega$ gives the same result. The real part of $e^{st}$ is also an even function of $\omega$, since that real part is $e^{\sigma}\cos(\omega t)$ and the cosine function is even. Hence the real parts of both $F(s)$ and $e^{st}$ are even functions of $\omega$, as is their product. A similar argument can be made for the product of the imaginary parts, since those imaginary parts are both odd functions of $\omega$, and the product of two odd functions is an even function (see the chapter-end problems and online solutions if you're not sure why the product of two even or two odd functions is even).

As you can see in the bottom portion of Fig. 2.4, the situation is quite different for the (imaginary) product of the real part of $F(s)$ with the imaginary part of $e^{st}$ and the (also imaginary) product of the imaginary part of $F(s)$ with the real part of $e^{st}$. In those cases, the curves for $-\omega$ and $+\omega$ have opposite signs, which means that the imaginary portion of the product $F(s)e^{st}$ will not contribute to any integration performed over the $\omega$ axis between symmetric limits (including $-\infty$ to $+\infty$).

But you should not fall into the trap of thinking that the imaginary portion of $F(s)$ cannot contribute to the integration performed in the inverse Laplace transform leading to a purely real time-domain function. It clearly does contribute to the real time-domain function $f(t)$ when multiplied by the imaginary part of $e^{st}$, and that contribution for a specific value of $\omega$ is indicated by the dashed curve in the middle portion of Fig. 2.4.

You should also note that the product of the real parts of $F(s)$ and $e^{st}$ involves the cosine function, while the product of the imaginary parts involves the sine function. That is very important in the synthesis of $f(t)$ by the inverse Laplace transform because the relative sizes of these terms, which is set by the relative size of the real and imaginary parts of $F(s)$, determines "how much cosine" and "how much sine" is present in the synthesized function. And as you will see in the other examples in this section, the relative amount of cosine and sine functions impacts the location in time of events such as zero crossings, peaks, and valleys of the synthesized function $f(t)$.

Once you understand the role of the real and imaginary parts of $F(s)$ at a single angular frequency $\omega$ in the inverse Laplace transform, you should

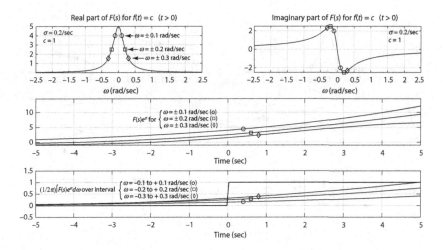

Figure 2.5 Three low-frequency components of $F(s) = c/s$.

be ready to appreciate how the time-domain function $f(t)$ is synthesized by integrating the product of the complex functions $F(s)$ and $e^{st}$ over $\omega$ for a given value of $\sigma$. The mathematical statement of the inverse Laplace transform is given in Chapter 1 as

$$f(t) = \frac{1}{2\pi i} \int_{\sigma-i\infty}^{\sigma+i\infty} F(s)e^{st} ds, \qquad (1.25)$$

in which the integral is taken over the entire $\omega$ axis, from $\omega = -\infty$ to $\omega = +\infty$.

To see how the sinusoidal frequency components weighted by $F(s)$ combine to produce the time-domain function $f(t)$, the next three figures illustrate the effect of limiting the integral in the inverse Laplace transform to various ranges of $\omega$. Doing that means that $f(t)$ is given by the equation

$$f(t) = \frac{1}{2\pi} \int_{\sigma-i\omega}^{\sigma+i\omega} F(s)e^{(\sigma+i\omega)t} d\omega, \qquad (2.7)$$

so the contributions to $f(t)$ come from the range of angular frequencies between $-\omega$ and $+\omega$. Note that integrating over a range of $\omega$ along a line of fixed $\sigma$ in the complex plane allows the substitution $d\omega = ds/i$, as described in Section 1.6.

You can see the result of limiting the integration range to low frequencies (small $\omega$) in Fig. 2.5, in which three small values of $\omega$ have been selected; in this figure, those values are $\omega = \pm 0.1$ rad/sec, $\omega = \pm 0.2$ rad/sec, and $\omega = \pm 0.3$ rad/sec.

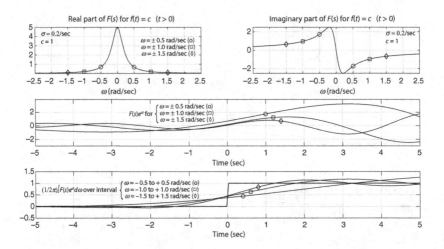

Figure 2.6 Three mid-frequency components of $F(s) = c/s$.

The top portion of this figure shows how much of the $s$-domain function $F(s)$ is encompassed by these values of $\omega$, and the middle portion shows a time-domain plot of each of these three low-frequency components (that is, it shows the product $F(s)e^{st}$ for each of these three values of $\omega$). Note that each of these components makes small contributions to the amplitude for $t < 0$ and larger contributions at times $t > 0$, as required for the time-domain function $f(t) = 0$ for $t < 0$ and $f(t) = c$ for $t > 0$.

Now look at the bottom portion of Fig. 2.5, which shows the results of integrating over the angular frequency ranges bounded by each of these three values of $\omega$. As you can see, these slowly changing low-frequency components cannot produce the sudden amplitude step at $t = 0$ required for $f(t)$, which is shown by the dark line in this plot. This clearly indicates that additional higher-frequency (faster-changing) components are needed to synthesize $f(t)$.

The effects of extending the angular frequency limits to $\omega = \pm 0.5$ rad/sec, $\pm 1.0$ rad/sec, and $\pm 1.5$ rad/sec are shown in Fig. 2.6. In the middle portion of this figure, you can see that, unlike the low-frequency components shown in the previous figure, these middle-frequency components "roll over" (that is, the slope changes sign) within the $\pm 5$-second time window shown on these plots. For times $t < 0$, you can see the effect described earlier in this section, in which the contributions from the imaginary portion of $F(s)$ multiplied by sine functions combine with the contributions from the real portion of $F(s)$

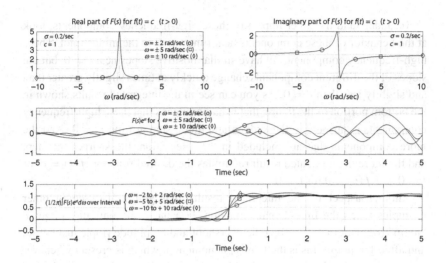

Figure 2.7  Three high-frequency components of $F(s) = c/s$.

multiplied by cosine functions. As shown in the bottom portion of this figure, that combination drives the integrated result toward zero for negative times.

For times $t > 0$, the contributions from the real and imaginary portions of $F(s)$ tend to increase rather than decrease the integrated result. But as shown in the middle plot, the peaks and valleys of these frequency components occur at different times due to the different values of $\omega$ in the $e^{st} = e^{(\sigma + i\omega)t}$ term, and that tends to smooth out the fluctuations in amplitude of the individual components. The integrated result shown in the bottom portion of the figure illustrates how this smoothing effect helps produce a constant amplitude near unity for $t > 0$, as required for the time-domain function $f(t) = c = 1$ for $t > 0$. This graph also shows that the same smoothing effect helps produce a constant amplitude near zero for $t < 0$.

The contributions of even higher-frequency components ($\omega = \pm 2$ rad/sec, $\pm 5$ rad/sec, and $\pm 10$ rad/sec) are shown in Fig. 2.7. The middle portion of this figure shows that these components go through several cycles in the time window of $\pm 5$ seconds, and the effect of those variations helps synthesize the function $f(t) = c$ in two important ways. Firstly, the different locations of the peaks and valleys of these rapidly varying components continues the smoothing process discussed above. That drives the integrated result closer to the desired function, as you can see in the bottom portion of this figure.

To understand the other impact of these high-frequency components, look at the behavior of these components near time $t = 0$ in the middle plot. These high-frequency components all have similar positive slopes near $t = 0$, but the slopes of the different components change quickly at times slightly greater than and slightly less than $t = 0$. As you can see in the integrated results shown in the bottom portion of the figure, that behavior means that these high-frequency components contribute to the sudden amplitude step at time $t = 0$. And the more components that are included in the integration, the sharper the step, and the more the integrated result resembles the desired function $f(t) = c$ for $t > 0$ and $f(t) = 0$ for $t < 0$.

Another aspect of the integrated results that you'll also see in other examples when the time-domain function $f(t)$ has a discontinuity or finite "jump" (often at time $t = 0$) is the overshoot and amplitude ripple just before and after the jump. This is the Gibbs phenomenon, which is present whenever you synthesize a discontinuous function using a finite number of continuous sinusoidal functions. Adding more high-frequency components reduces the period of the Gibbs ripples, so the time extent of a ripple of a given amplitude will get shorter, but the overshoot will remain approximately 9% of the size of the jump.

The important concept demonstrated by the last three figures is that the sinusoidal frequency components, weighted by the real and imaginary parts of the $s$-domain function $F(s)$ and integrated over $\omega$, do approach the time-domain function $f(t)$ of which $F(s)$ is the Laplace transform. And the more frequency components you include in the mix, the closer the function synthesized by the inverse Laplace transform gets to $f(t)$.

## 2.2 Exponential Functions

Almost as straightforward as a constant time-domain function is an exponential function of the form

$$f(t) = e^{at}, \tag{2.8}$$

in which $a$ represents a real constant. An example of this function with $a = 1/\text{sec}$ is shown in Fig. 2.8a.

When the constant $a$ equals zero, this exponential function is identical to the constant function $F(s) = c$ with $c = 1$, so the Laplace transform of $f(t) = e^{at}$ should reduce to $F(s) = 1/s$ within the region of convergence. But if the

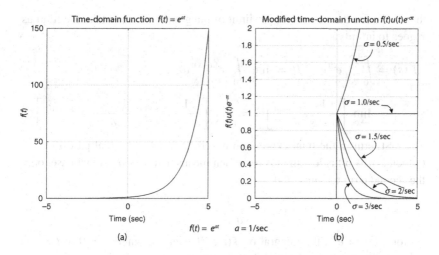

Figure 2.8 (a) Exponential time-domain function $f(t)$ and (b) modified $f(t)$.

constant $a$ in the function $f(t) = e^{at}$ is greater than zero, this time-domain function grows exponentially as time increases, and the Laplace-transform integral will not converge unless the decrease of the multiplicative term $e^{-\sigma t}$ in the modified function $f(t)u(t)e^{-\sigma t}$ is faster than the increase of $e^{at}$ over time.

The effect of the $e^{-\sigma t}$ term, as well as the effect of the step function $u(t)$ for several values of $\sigma$, is shown in Fig. 2.8b. As you can see, with $\sigma = 0.5/\mathrm{sec}$, the modified function $f(t)u(t)e^{-\sigma t}$ still increases over time, which means that the Laplace-transform integral will not converge in this case. At $\sigma = 1/\mathrm{sec}$, the decrease in $e^{-\sigma t}$ just compensates for the increase in the function $e^{at}$ over time, so a slightly larger value of $\sigma$ is needed for the integral to converge. Three examples of this ($\sigma$ values of $1.5/\mathrm{sec}$, $2/\mathrm{sec}$, and $3/\mathrm{sec}$) are shown in this figure, which demonstrates that larger values of $\sigma$ produce a faster decrease of the modified function over time due to the higher number of $1/e$ steps per second.

Here is the mathematical statement of the unilateral Laplace transform of an exponential function:

$$F(s) = \int_{0^-}^{+\infty} f(t)e^{-st}\,dt = \int_0^{+\infty} e^{at}e^{-st}\,dt.$$

As in the case of the constant function $f(t) = c$, the lower limit in the second integral can be set to zero since this time-domain function is continuous across time $t = 0$. In this case, this improper integral converges as long as $\sigma$ is greater than $a$. To see that, set the upper limit of integration to $\tau$ and take the limit as $\tau$ goes to infinity:

$$F(s) = \int_0^{+\infty} e^{at} e^{-st} dt = \lim_{\tau \to \infty} \int_0^{\tau} e^{(a-s)t} dt$$

$$= \lim_{\tau \to \infty} \left[ \frac{1}{a-s} e^{(a-s)t} \right]_0^{\tau} = \lim_{\tau \to \infty} \left[ \frac{1}{a-s} e^{(a-s)\tau} - \frac{1}{a-s} e^{(a-s)(0)} \right].$$

The first term in the final expression is zero as long as the real part of $a - s$ (which is $a - \sigma$) is less than zero, which means that $\sigma > a$. In that case, only the second term remains, and

$$F(s) = -\frac{1}{a-s} e^0 = \frac{1}{s-a}. \tag{2.9}$$

So for $f(t) = e^{at}$, the integral of $f(t)e^{-st}$ over the range $t = 0$ to $t = \infty$ converges to a value of $1/(s-a)$ as long as $s > a$:

$$F(s) = \frac{1}{s-a} \qquad\qquad \text{ROC: } s > a. \tag{2.10}$$

The behavior of $F(s)$ as a function of $\sigma$ and $\omega$ is similar to that of $F(s)$ in the previous example, but note that the regions of convergence are different: $\sigma > 0$ for $f(t) = c$ and $\sigma > a$ for $f(t) = e^{at}$. A plot of $F(s) = 1/(s-a)$ at constant $\sigma = 2/\text{sec}$ looks identical to the plot of $F(s) = 1/s$ (Fig. 2.3) at $\sigma = 1/\text{sec}$, so instead of repeating that plot, Fig. 2.9 shows the effect of taking

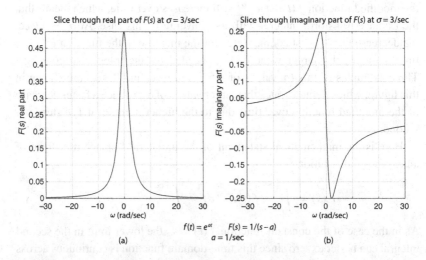

Figure 2.9 (a) Real and (b) imaginary parts of a slice through $F(s) = 1/(s-a)$ at $\sigma = 3/\text{sec}$ and $a = 1/\text{sec}$.

a slice at a larger value of $\sigma$ (which is $\sigma = 3$/sec in this case). As you can see, both the real and the imaginary parts of this slice through $F(s)$ have the same general shape as a slice taken at a smaller value of $\sigma$, although the spread of values near $\omega = 0$ is somewhat greater in this case.

That greater spread (meaning larger amplitudes of frequency components on either side of $\omega = 0$) can be understood by looking back on the plot of the modified time-domain function $f(t)u(t)e^{-\sigma t}$ in Fig. 2.8b. For $\sigma = 3$/sec, the decrease in amplitude of the modified function over time is steeper, which means that higher-frequency components with greater amplitudes must be mixed in to produce this modified time-domain function. That logic applies to the cosine components shown in Fig. 2.9a as well as the sine components shown in Fig. 2.9b, since both sets of components are needed to smooth out the ripples in the synthesized function $f(t)u(t)$ for $t > 0$ and to ensure that this function has zero value for $t < 0$.

To understand how the inverse Laplace transform of the complex-frequency function $F(s) = \frac{1}{s-a}$ gives the time-domain function $f(t) = e^{at}$ for $t > 0$, take a look at Fig. 2.10. In the middle portion of this figure, you can see one relatively low-frequency component ($\omega = \pm 0.2$ rad/sec), one mid-frequency component ($\omega = \pm 3$ rad/sec), and one high-frequency component ($\omega = \pm 10$ rad/sec) of $F(s)$ for $a = 0.2$/sec. The bottom portion of this figure shows the result of integrating $F(s)$ over the three frequency ranges bounded

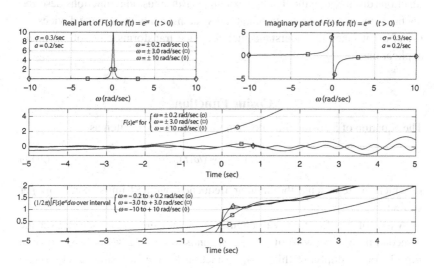

Figure 2.10 Low-, medium-, and high-frequency components of $F(s) = \frac{1}{s-a}$ for $f(t) = e^{at}$.

by these values; in this case the integration is performed along the $\sigma = 0.3/\text{sec}$ line, just within the region of convergence $(\sigma > a)$ in the complex plane for the Laplace transform of $f(t) = e^{at}$.

As you can see, the roles of these three frequency components are similar to those in the previous example of $f(t) = c$ for $t > 0$. Just as in that case, the low-frequency component establishes the small-amplitude baseline for times $t < 0$, along with the larger (and in this case exponentially increasing) amplitude for times $t > 0$. The mid-frequency components further reduce the amplitude of the integrated result for negative time while also increasing the amplitude for positive time toward the desired function $e^{at}$. And, as expected, the quickly changing high-frequency components contribute to the amplitude step at $t = 0$ and help smooth out the ripples in the integrated result. Extending the integration to a wider frequency range will further sharpen the step and drive the output of the inverse Laplace transform even closer to $f(t) = e^{at}$ for $t > 0$ (shown as a dark line in this plot).

## 2.3 Sinusoidal Functions

The sinusoidal time-domain functions $\cos(\omega t)$ and $\sin(\omega t)$ are the subject of this section, and as you will see, the Laplace transforms of these functions are somewhat more complicated than those of the constant and exponential functions discussed in the first two examples. But sinusoidal functions describe the behavior of a wide variety of natural systems and human-made devices, so the effort it takes to understand their Laplace transforms pays off in many applications.

### Cosine Functions

The equation of a cosine time-domain function can be written as

$$f(t) = \cos(\omega_1 t), \tag{2.11}$$

in which $\omega_1$ represents the angular frequency, with SI units of radians per second.

A plot of this function with angular frequency $\omega_1 = 20$ rad/sec over the 4-second time window from $t = -2$ seconds to $t = +2$ seconds is shown in Fig. 2.11a. Multiplying this cosine function by the unit-step function $u(t)$ and by the real exponential $e^{-\sigma t}$ results in the modified function shown in Fig. 2.11b, in which $\sigma = 1/\text{sec}$.

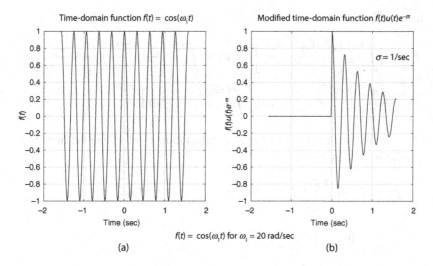

Figure 2.11 (a) Cosine time-domain function $f(t)$ and (b) modified function $f(t)u(t)e^{-\sigma t}$ for $\sigma = 1/\text{sec}$.

If you have read Section 1.5, you may recall that the Fourier transform of this modified function does converge as long as $\sigma > 0$, with the result

$$F(\omega) = \frac{s}{s^2 + \omega_1^2} = \frac{\sigma + i\omega}{(\sigma + i\omega)^2 + \omega_1^2}. \qquad (1.22)$$

As described in Chapter 1, the unilateral Laplace transform of $f(t)$ gives the same result as the Fourier transform of the modified function $f(t)u(t)e^{-\sigma t}$, and here's how that transform looks when written in terms of the generalized frequency $s = \sigma + i\omega$:

$$F(s) = \int_{0^-}^{+\infty} f(t)e^{-st}dt = \int_{0}^{+\infty} \cos(\omega_1 t)e^{-st}dt, \qquad (2.12)$$

in which the lower limit of 0 gives the same result as a lower limit of $0^-$ because the cosine function is continuous across time $t = 0$. Using the inverse Euler relation

$$\cos(\omega_1 t) = \frac{e^{i\omega_1 t} + e^{-i\omega_1 t}}{2} \qquad (2.13)$$

makes this

$$F(s) = \int_0^{+\infty} \left[ \frac{e^{i\omega_1 t} + e^{-i\omega_1 t}}{2} \right] e^{-st} dt = \int_0^{+\infty} \left[ \frac{e^{(i\omega_1 - s)t} + e^{(-i\omega_1 - s)t}}{2} \right] dt$$

$$= \frac{1}{2} \left[ \int_0^{+\infty} e^{-(s-i\omega_1)t} dt + \int_0^{+\infty} e^{-(s+i\omega_1)t} dt \right].$$

As before, calling the upper limit of integration $\tau$ and taking the limit as $\tau$ goes to infinity gives

$$F(s) = \lim_{\tau \to \infty} \frac{1}{2} \left[ \int_0^\tau e^{-(s-i\omega_1)t} dt + \int_0^\tau e^{-(s+i\omega_1)t} dt \right]$$

$$= \lim_{\tau \to \infty} \frac{1}{2} \left[ \frac{1}{-(s-i\omega_1)} e^{-(s-i\omega_1)t} \right]\Big|_0^\tau$$

$$+ \lim_{\tau \to \infty} \frac{1}{2} \left[ \frac{1}{-(s+i\omega_1)} e^{-(s+i\omega_1)t} \right]\Big|_0^\tau$$

$$= \lim_{\tau \to \infty} \frac{1}{2} \left[ \frac{1}{-(s-i\omega_1)} e^{-(s-i\omega_1)\tau} - \frac{1}{-(s-i\omega_1)} e^{-(s-i\omega_1)0} \right]$$

$$+ \lim_{\tau \to \infty} \frac{1}{2} \left[ \frac{1}{-(s+i\omega_1)} e^{-(s+i\omega_1)\tau} - \frac{1}{-(s+i\omega_1)} e^{-(s+i\omega_1)0} \right].$$

Inside each of the large square brackets, the exponential terms in which $\tau$ appears will approach zero as $\tau \to \infty$ as long as $\sigma > 0$ (if you're concerned about the imaginary parts of those terms, remember that terms of the form $e^{i\omega t}$ simply oscillate between $+1$ and $-1$ no matter how large $t$ becomes). Thus

$$F(s) = \frac{1}{2} \left[ 0 - \frac{1}{-(s-i\omega_1)} e^{-(s-i\omega_1)0} \right] + \frac{1}{2} \left[ 0 - \frac{1}{-(s+i\omega_1)} e^{-(s+i\omega_1)0} \right]$$

$$= \frac{1}{2} \left[ \frac{1}{(s-i\omega_1)} + \frac{1}{(s+i\omega_1)} \right],$$

which rationalizes to

$$F(s) = \frac{1}{2} \left[ \frac{s+i\omega_1}{(s-i\omega_1)(s+i\omega_1)} + \frac{s-i\omega_1}{(s+i\omega_1)(s-i\omega_1)} \right]$$

$$= \frac{1}{2} \left[ \frac{s+i\omega_1 + s - i\omega_1}{s^2 + \omega_1^2} \right] = \frac{1}{2} \left[ \frac{2s}{s^2 + \omega_1^2} \right] = \frac{s}{s^2 + \omega_1^2}.$$

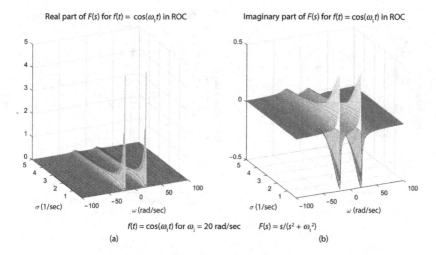

Figure 2.12 (a) Real and (b) imaginary parts of $F(s)$ for cosine $f(t)$ in region of convergence.

So in this case the integral of $f(t)e^{-st}$ over the range $t = 0$ to $t = \infty$ converges to a value of $\frac{s}{s^2+\omega_1^2}$ as long as $s > 0$. Thus the unilateral Laplace transform of the pure cosine function $f(t) = \cos(\omega_1 t)$ is

$$F(s) = \frac{s}{s^2 + \omega_1^2} \qquad \text{ROC: } s > 0, \qquad (2.14)$$

which is identical to the result of the Fourier transform of the modified function $f(t)u(t)e^{-\sigma t}$ shown in Chapter 1.

Plots of the real and imaginary parts of $F(s)$ for this case are shown in Fig. 2.12. As in previous examples, the real part of $F(s)$ is considerably larger than the imaginary part. That shouldn't be too surprising, since the real part of $F(s)$ tells you the amount of the $\cos(\omega_1 t)$ function that is present in the modified time-domain function $f(t)$, and in this example the unmodified function $f(t)$ is a pure cosine function (that is, it was a pure cosine function before multiplication by the unit-step function $u(t)$ and the exponential decay function $e^{-\sigma t}$).

Recall from Chapter 1 that the frequency spectrum $F(\omega)$ of a pure cosine function consists of two positive delta functions, one at $+\omega_1$ and the other at $-\omega_1$, representing two counter-rotating phasors (described by $e^{i\omega_1 t}$ and $e^{-i\omega_1 t}$). The combination of those two phasors produces a purely real function; the imaginary part of $F(\omega)$ is zero because no sine components are needed to make up a pure cosine function. But as described in the previous examples,

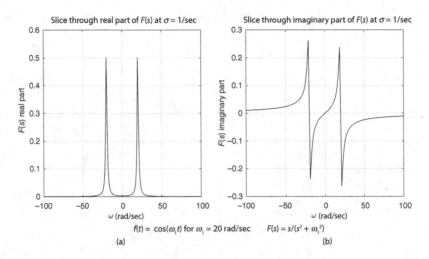

Figure 2.13 (a) Real and (b) imaginary parts of slice through F(s) at $\sigma = 1/\text{sec}$ for $f(t) = \cos(\omega_1 t)$ with $\omega_1 = 20$ rad/sec.

in the unilateral Laplace transform the modifications to the pure cosine function mean that additional frequency components are needed to produce a time-domain function that is zero for all negative time ($t < 0$) and that has decreasing amplitude for positive time ($t > 0$). You can see those additional frequency components as the spreading out of the peaks at $-\omega_1$ and $+\omega_1$ in the slice through $F(s)$ shown in Fig. 2.13, taken at $\sigma = 1/\text{sec}$.

The amplitudes of the additional frequency components near the two positive peaks shown in the slice through the real part of $F(s)$ (Fig. 2.13a) are all positive, so these components are positive cosine basis functions, which are just what are needed to produce amplitude roll-off caused by the $e^{-\sigma t}$ term in the modified time-domain function $f(t)u(t)e^{-\sigma t}$. But as you can see in Fig. 2.13b, the imaginary parts of the frequency components have a phase inversion of 180 degrees (that is, a change in sign) at $\omega = -\omega_1$ and at $\omega = +\omega_1$. As you'll see in the discussion of the inverse Laplace transform of $F(s) = s/(s^2 + \omega_1^2)$ later in this section, this sign change allows the sine components at frequencies $-\omega_1 < \omega < 0$ and $0 < \omega < \omega_1$ to combine with the cosine components to reduce the amplitude of the integrated result at negative time, while the higher-frequency sine components serve to smooth out the ripples in the synthesized function produced by the inverse Laplace transform.

Looking closely at the slice through the imaginary part of $F(s)$, you may notice that the peaks closer to the middle of the plot (nearer to $\omega = 0$) have slightly smaller amplitude than the peaks farther out. That occurs because of

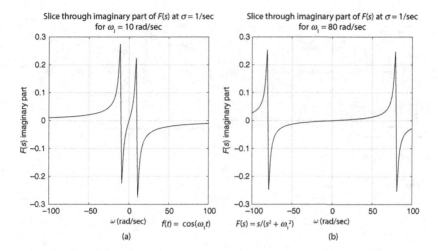

Figure 2.14 Imaginary part of a slice through $F(s)$ at $\sigma = 1/\text{sec}$ for $f(t) = \cos(\omega_1 t)$ with (a) $\omega_1 = 10$ rad/sec and (b) $\omega_1 = 80$ rad/sec.

the influence of the peaks on the opposite side of $\omega = 0$. Since those peaks are spread out in both directions (that is, toward higher as well as lower frequencies), their "tails" may spread across $\omega = 0$ to the other side of the spectrum, and that slightly changes the amplitude of the peaks on the opposite side. The closer the peaks are to $\omega = 0$, the higher the amplitude of those tails on the opposite side, so the effect is greater for smaller $\omega_1$, and it is less for greater values of $\omega_1$, as you can see in Fig. 2.14 for $\omega_1 = 10$ rad/sec and 80 rad/sec.

Notice, however, that this effect "goes both ways" – that is, the tails of the peaks on the left side of the spectrum ($\omega < 0$) influence the amplitude of the peaks on the right side of the spectrum ($\omega > 0$) in exactly the same way that the peaks on the right side influence the amplitudes on the left side. As a result, the imaginary part of the frequency spectrum retains the antisymmetric form about $\omega = 0$ that is needed for the sine components to combine with the cosine components to reduce the amplitude of the integrated result for $t < 0$ and to add to the cosine components for $t > 0$ to produce the desired time-domain function.

The process through which the inverse Laplace transform produces the time-domain function $f(t) = \cos(\omega_1 t)$ can be understood by considering the behavior of the low-, medium-, and high-frequency components that are weighted by the real and imaginary parts of $F(s)$ and integrated in the inverse transform. In this context "low frequency" refers to components with

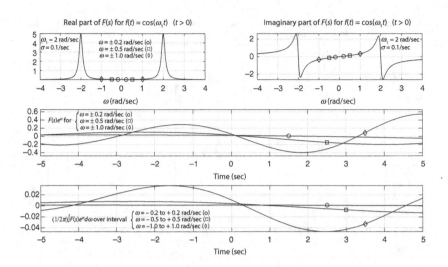

Figure 2.15  Three low-frequency components of $F(s)$ for $f(t) = \cos(\omega_1 t)$.

frequencies smaller than $\omega_1$, "medium frequency" refers to components with frequencies that are close to $\omega_1$, and "high frequency" refers to components with frequencies greater than $\omega_1$.

In Fig. 2.15, you can see three of the low-frequency components of $F(s) = s/(s^2 + \omega_1^2)$ for $\omega_1 = 2$ rad/sec with $\sigma = 0.1$/sec (just inside the region of convergence). As shown in the middle portion of this figure, these three components, at angular frequencies of $\pm 0.2$ rad/sec, $\pm 0.5$ rad/sec, and $\pm 1.0$ rad/sec, have relatively small amplitudes because $F(s)$ is small at these frequencies. The integrated results shown in the bottom portion of this figure are very small relative to the time-domain function $f(t) = \cos(\omega_1 t)$ (which is not shown in this plot because its amplitude is so much greater than the integrated results over these narrow frequency ranges). But if you look carefully at the shape of these low-frequency components, you can see that they have the shape of a negative sine function – that is, they are negative for the first half-cycle to the right of $t = 0$ and positive for the first half-cycle to the left of $t = 0$. That shape derives largely from the contribution from the imaginary portion of $F(s)$, and as described earlier in this section, the sign inversion that occurs at $\omega = \omega_1$ for the imaginary part of $F(s)$ changes that shape to a positive sine function for higher frequencies $\omega > \omega_1$ and $\omega < -\omega_1$.

You can see the change in behavior of the components at frequencies just below and just above $\omega_1$ in Fig. 2.16. The middle portion of this figure shows the components at frequencies of $\omega = \pm 1.9$ rad/sec, $\pm 2.0$ rad/sec

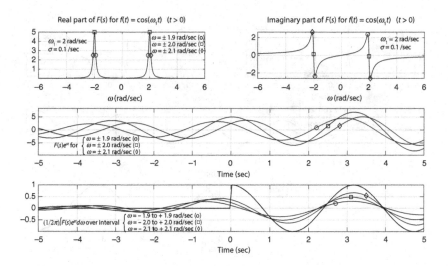

Figure 2.16 Three medium-frequency components of $F(s)$ for $f(t) = \cos(\omega_1 t)$.

(which is $\omega_1$), and $\pm 2.1$ rad/sec. All of these components have positive amplitude at time $t = 0$, as you might expect for the components of the time-domain function $f(t) = \cos(\omega_1 t)$. But notice that the component with frequency $\omega = \pm 1.9$ rad/sec, just below $\omega_1$, is shifted to the left in time while the component with frequency $\omega = \pm 2.1$ rad/sec, just above $\omega_1$, is shifted to the right in time. That is caused by the contribution of the imaginary potion of $F(s)$, which changes sign at frequency $\omega = \omega_1$ and $\omega = -\omega_1$. And when $\omega$ exactly equals $\pm \omega_1$, the imaginary portion of $F(s)$ is zero, and the shape of that component is a pure cosine function, reaching a peak at time $t = 0$.

Now consider the results of integrating over these medium-frequency ranges shown in the bottom portion of Fig. 2.16. The general shape of these integrated results is approaching the shape of the cosine function $f(t) = \cos(\omega_1 t)$, and although the amplitudes are smaller for times $t < 0$ than for times $t > 0$, driving the results closer to zero for all negative times requires the contributions of higher-frequency components.

Three of those higher-frequency components are shown in Fig. 2.17. The components shown in the middle portion of this figure have angular frequencies of $\omega = \pm 3$ rad/sec, $\pm 5$ rad/sec, and $\pm 10$ rad/sec, and as you can see, all three of these components have the shape of a positive sine function. That is largely due to the contributions from the imaginary part of $F(s)$, which outweigh the contributions from the real part of $F(s)$ at these frequencies.

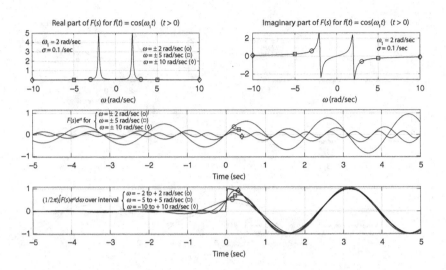

Figure 2.17  Three high-frequency components of $F(s)$ for $f(t) = \cos(\omega_1 t)$.

Notice also that the amplitude of these components is negative for the first half-cycle to the left of $t = 0$; this helps drive the integrated result toward zero for negative time, as required for $f(t) = 0$ for $t < 0$. But in the first half-cycle to right of $t = 0$, these components help form the step required for the time-domain function to go from zero for $t < 0$ to one for $t = 0$, since the cosine function has a value of one at that time. As you may have anticipated if you've read the other examples in this chapter, including more high-frequency components into the mix by integrating the product $F(s)e^{st}$ over a wider range of $\omega$ drives the integrated response closer to the time-domain function $f(t)$ for which $F(s)$ is the Laplace transform.

## Sine Functions

The mathematics of the Laplace transform of a pure sine function such as $f(t) = \sin(\omega_1 t)$ has a lot in common with that of the pure cosine function, but the resulting $s$-domain function $F(s)$ has an important difference. To see how that difference comes about, and to understand why it's important, start with the time-domain function

$$f(t) = \sin(\omega_1 t) \tag{2.15}$$

shown in Fig. 2.18a for $\omega_1 = 20$ rad/sec and apply the modifications of multiplying $f(t)$ by $e^{-\sigma t}$ and by $u(t)$. That gives the modified function shown in Fig. 2.18b for $\sigma = 1/\text{sec}$.

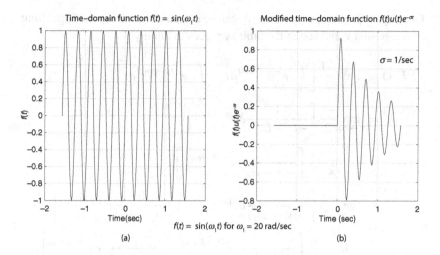

Figure 2.18 (a) Sine time-domain function $f(t)$ and (b) modified $f(t)$.

To find $F(s)$, follow a similar path to that shown above for a cosine time-domain function. Start by writing the unilateral Laplace transform as

$$F(s) = \int_{0^-}^{+\infty} f(t)e^{-st}dt = \int_0^{+\infty} \sin(\omega_1 t)e^{-st}dt,$$

in which the lower limit in the second integral can be set to 0 since $f(t) = \sin(\omega_1 t)$ is continuous across time $t = 0$. Now apply the inverse Euler relation for sine:

$$\sin(\omega t) = \frac{e^{i\omega t} - e^{-i\omega t}}{2i}, \tag{2.16}$$

and the next steps should look familiar:

$$F(s) = \int_0^{+\infty} \left[\frac{e^{i\omega_1 t} - e^{-i\omega_1 t}}{2i}\right]e^{-st}dt = \int_0^{+\infty} \left[\frac{e^{(i\omega_1 - s)t} - e^{(-i\omega_1 - s)t}}{2i}\right]dt$$

$$= \frac{1}{2i}\left[\int_0^{+\infty} e^{(i\omega_1 - s)t}dt - \int_0^{+\infty} e^{(-i\omega_1 - s)t}dt\right]$$

$$= \frac{-i}{2}\left[\int_0^{+\infty} e^{-(s-i\omega_1)t}dt - \int_0^{+\infty} e^{-(s+i\omega_1)t}dt\right].$$

Once again these improper integrals can be evaluated by calling the upper limit of integration $\tau$ and taking the limit as $\tau$ goes to infinity:

$$
F(s) = \lim_{\tau \to \infty} \frac{-i}{2} \left[ \int_0^\tau e^{-(s-i\omega_1)t} \, dt - \int_0^\tau e^{-(s+i\omega_1)t} \, dt \right]
$$

$$
= \lim_{\tau \to \infty} \frac{-i}{2} \left[ \frac{1}{-(s-i\omega_1)} e^{-(s-i\omega_1)t} \right]\Big|_0^\tau
$$

$$
- \lim_{\tau \to \infty} \frac{-i}{2} \left[ \frac{1}{-(s+i\omega_1)} e^{-(s+i\omega_1)t} \right]\Big|_0^\tau
$$

$$
= \lim_{\tau \to \infty} \frac{-i}{2} \left[ \frac{1}{-(s-i\omega_1)} e^{-(s-i\omega_1)\tau} - \frac{1}{-(s-i\omega_1)} e^{-(s-i\omega_1)0} \right]
$$

$$
- \lim_{\tau \to \infty} \frac{-i}{2} \left[ \frac{1}{-(s+i\omega_1)} e^{-(s+i\omega_1)\tau} - \frac{1}{-(s+i\omega_1)} e^{-(s+i\omega_1)0} \right].
$$

Just as in the cosine case, inside each of the large square brackets the exponential terms in which $\tau$ appears will approach zero as $\tau \to \infty$ as long as $\sigma > 0$ (and, as before, you needn't worry about the imaginary parts of those terms because they simply oscillate between $+1$ and $-1$ no matter how large $t$ becomes). Hence

$$
F(s) = \frac{-i}{2} \left[ 0 - \frac{1}{-(s-i\omega_1)} e^{-(s-i\omega_1)0} \right] - \frac{-i}{2} \left[ 0 - \frac{1}{-(s+i\omega_1)} e^{-(s+i\omega_1)0} \right]
$$

$$
= \frac{-i}{2} \left[ \frac{1}{(s-i\omega_1)} - \frac{1}{(s+i\omega_1)} \right],
$$

which rationalizes to

$$
F(s) = \frac{-i}{2} \left[ \frac{s+i\omega_1}{(s-i\omega_1)(s+i\omega_1)} - \frac{s-i\omega_1}{(s+i\omega_1)(s-i\omega_1)} \right]
$$

$$
= \frac{-i}{2} \left[ \frac{s+i\omega_1 - s+i\omega_1}{s^2+\omega_1^2} \right] = \frac{-i}{2} \left[ \frac{2i\omega_1}{s^2+\omega_1^2} \right] = \frac{\omega_1}{s^2+\omega_1^2}.
$$

So for the time-domain function $f(t) = \sin(\omega_1 t)$ the integral of $f(t)e^{-st}$ over the range $t = 0$ to $t = \infty$ converges to a value of $\omega_1/(s^2 + \omega_1^2)$ as long as $s > 0$. Thus the unilateral Laplace transform of this pure sine function is

$$
F(s) = \frac{\omega_1}{s^2+\omega_1^2} \qquad\qquad \text{ROC: } s > 0. \qquad\qquad (2.17)
$$

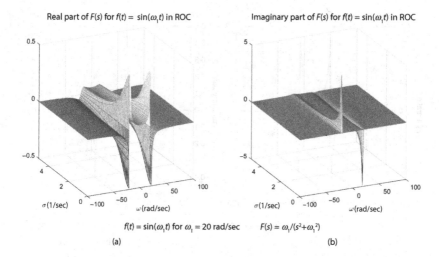

Real part of $F(s)$ for $f(t) = \sin(\omega_t t)$ in ROC        Imaginary part of $F(s)$ for $f(t) = \sin(\omega_t t)$ in ROC

$f(t) = \sin(\omega_t t)$ for $\omega_1 = 20$ rad/sec        $F(s) = \omega_t/(s^2+\omega_t^2)$

(a)        (b)

Figure 2.19 (a) Real and (b) imaginary parts of $F(s)$ for $f(t) = \sin(\omega_1 t)$ with $\omega_1 = 20$ rad/sec in ROC.

As you can see, the only difference between this expression for the Laplace transform result $F(s)$ of a sine function and $F(s)$ for a cosine function (Eq. 2.14) is that the numerator is $\omega_1$ in the sine case instead of $s$ for the cosine case. But even a quick glance at a plot of the real and imaginary parts of $F(s)$ for a time-domain sine function, such as Fig. 2.19, reveals substantial differences.

The first thing you may notice is that the imaginary part is larger, which you may have expected since the imaginary part tells you the amount of sine basis functions in the modified $f(t)$, and in this case $f(t)$ is a pure sine function. But you should also notice that the real portion of $F(s)$ is small but not zero, since the real part tells you the amount of cosine basis functions in the modified $f(t)$, and some cosine components are needed to combine with the sine components to reduce the amplitude of the integrated result for negative times ($t < 0$).

The detailed shape of the real and imaginary parts of $F(s)$ can be seen more easily by taking a slice through $F(s)$, such as that shown in Fig. 2.20, taken at $\sigma = 1/\text{sec}$. Figure 2.20b shows that the slice through the imaginary part of $F(s)$ resembles the plot of the imaginary part of the Fourier transform $F(\omega)$ of a pure sine wave (see Fig. 1.9), which consists of a positive delta function at $\omega = -\omega_1$ and a negative delta function at $\omega = \omega_1$. In this case, the imaginary part of $F(s)$ does have a positive peak at $\omega = -\omega_1$ and a negative peak at $\omega = +\omega_1$, but these peaks are neither infinitely tall nor infinitely narrow. And unlike the Fourier transform of a pure sine function, the Laplace

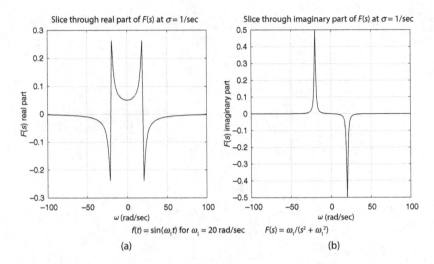

Figure 2.20 (a) Real and (b) imaginary parts of a slice through $F(s)$ at $\sigma = 1/\text{sec}$ for $f(t) = \sin(\omega_1 t)$ with $\omega_1 = 20$ rad/sec.

transform of $f(t) = \sin(\omega_1 t)$ has nonzero real components. As described in previous examples, that's because additional frequency components (both sines and cosines) are needed to produce the modified time-domain function $f(t)u(t)e^{-\sigma t}$, which has zero amplitude for all negative time and a decreasing envelope for positive time.

As shown in Fig. 2.20b, the slice through the imaginary part of $F(s)$ is antisymmetric, with positive values near the peak at $-\omega_1$ and negative values near the peak at $+\omega_1$. For negative angular frequencies ($\omega < 0$), multiplying the positive values of the imaginary part of $F(s)$ by the sine basis functions $\sin(-\omega t) = -\sin(\omega t)$ results in negative sine components. Likewise, at positive angular frequencies ($\omega > 0$), multiplying the negative values of the imaginary part of $F(s)$ by the sine basis functions $\sin(\omega t)$ also results in negative sine components. But since these components represent the imaginary part of $F(s)$, a factor of $i \times i = -1$ arises when these components are multiplied by the factor $e^{st} = e^{(\sigma + i\omega)t} = e^{\sigma t}(\cos(\omega t) + i \sin(\omega t))$ in the process of synthesizing $f(t)$ in the inverse Laplace transform. You can see the details of the process later in this section.

Similar logic can be applied to the real part of $F(s)$ shown in Fig. 2.20a. In this case, the component values are symmetric about $\omega = 0$, and since $\cos(\omega t) = \cos(-\omega t)$, the product of the real part of $F(s)$ and the cosine basis functions from the left side of the spectrum ($\omega < 0$) have the same sign as the products from the right side of the spectrum ($\omega > 0$). But as in the case of

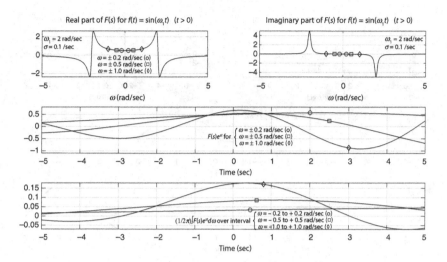

Figure 2.21  Three low-frequency components of $F(s)$ for $f(t) = \sin(\omega_1 t)$.

the imaginary part of $F(s)$ for $f(t) = \cos(\omega_1 t)$, the components of the real part of $F(s)$ for $f(t) = \sin(\omega_1 t)$ undergo a sign inversion at $\omega = -\omega_1$ and at $\omega = +\omega_1$. In this case, the products of the $F(s)$ values and the cosine basis functions are positive for frequencies $-\omega_1 < \omega < 0$ and $0 < \omega < +\omega_1$. Since there's no $i \times i$ factor for cosine components, these components add as positive cosine functions in the time-domain function $f(t)$ synthesized by the inverse Laplace transform, combining with the contributions of the sine components to reduce the amplitude of the integrated result for times $t < 0$. And as in the previous case, the higher-frequency components serve to smooth out the ripples in the synthesized time-domain function.

The final three figures of this section illustrate the process through which the time-domain function $f(t) = \sin(\omega_1 t)$ is produced by the inverse Laplace transform of the complex-frequency function $F(s) = \omega_1/(s^2 + \omega_1^2)$. As in the case of a cosine time-domain function, one approach to understanding this process is to look at several of the low-, medium-, and high-frequency components that contribute to the inverse transform.

Figure 2.21 shows three of the low-frequency components of $F(s)e^{st}$ with $\omega_1 = 2$ rad/sec and with $\sigma = 0.1/\text{sec}$. At these low angular frequencies ($\omega = \pm 0.2$ rad/sec, $\pm 0.5$ rad/sec, and $\pm 1.0$ rad/sec), the magnitude of the real part of $F(s)$ is considerably larger than the magnitude of the imaginary part, so the dominant contributions are cosine functions, as you can see in the middle portion of this figure. As mentioned above, these contributions are positive

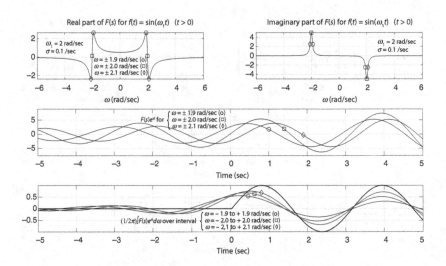

Figure 2.22 Three medium-frequency components of $F(s)$ for $f(t) = \sin(\omega_1 t)$.

cosine functions, since the real part of $F(s)$ is positive for $-\omega_1 < \omega < 0$ and $0 < \omega < +\omega_1$ and since $\cos(-\omega t) = \cos(\omega t)$. Over these narrow frequency ranges the integrated results shown in the bottom portion of this figure bear little resemblance to the desired function $f(t) = \sin(\omega_1 t)$ and are very small, which is why $f(t) = \sin(\omega_1 t)$ is not visible in the figure.

Adding more frequencies into the mix moves the integrated result toward $f(t)$, as you can see in Fig. 2.22. At frequencies of $\omega = \pm 1.9$ rad/sec (slightly smaller in magnitude than $\omega_1$) and $\omega = \pm 2.1$ rad/sec (slightly larger in magnitude than $\omega_1$), the real part of $F(s)$ has a sign inversion, so the sign of the cosine components changes at $\omega = \pm \omega_1$. At those two frequencies, the real part of $F(s)$ is zero, so the component has the shape of a pure sine wave, as shown in the middle portion of this figure.

That sine wave is positive because the contributions from both sides of the spectrum of $F(s)$ are positive. For the negative side, the imaginary part of $F(s)$ is positive near $\omega = -\omega_1$ and $i^2 \sin(-\omega t) = -i^2 \sin(\omega t) = \sin(\omega t)$, so the real product of the imaginary parts of $F(s)$ and $e^{(\sigma + i\omega)t}$ is positive. For the positive side, the imaginary part of $F(s)$ is negative near $\omega = +\omega_1$ and $i^2 \sin(+\omega t) = -\sin(\omega t)$, so the real product of the (negative) imaginary parts of $F(s)$ and $e^{(\sigma + i\omega)t}$ is positive. The bottom portion of this figure shows that adding these positive sine functions into the mix drives the integrated result closer to the desired time-domain function $f(t) = \sin(\omega_1 t)$ for time $t > 0$.

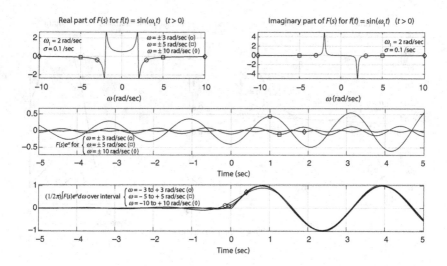

Figure 2.23  Three high-frequency components of $F(s)$ for $f(t) = \sin(\omega_1 t)$.

The results of integrating over wider frequency ranges are shown in Fig. 2.23. At these high frequencies of $\omega = \pm 3$ rad/sec, $\pm 5$ rad/sec, and $\pm 10$ rad/sec, the real part of $F(s)$ is larger than the imaginary part, so the dominant contributions come from cosine functions. As shown in the middle portion of this figure, these are negative cosine functions, since the real part of $F(s)$ is negative at these frequencies and since $\cos(-\omega t) = \cos(\omega t)$.

Note also that the imaginary part of $F(s)$ has positive small amplitude for negative frequencies and negative small amplitude for positive frequencies, both of which contribute positive sine functions to the integrated result. These components serve to smooth out the ripples in the integrated result and to push the amplitude toward zero for negative time and toward $\sin(\omega_1 t)$ for positive time, as needed to synthesize the time-domain function $f(t) = 0$ for $t < 0$ and $f(t) = \sin(\omega_1 t)$ for $t > 0$.

## 2.4  $t^n$ Functions

The shape of the time-domain function $f(t) = t^n$ varies significantly depending on the value of $n$, from a constant for $n = 0$, to a slanted line for $n = 1$, a parabola for $n = 2$, a cubic for $n = 3$, and so forth, with increasing curvature as $n$ increases. You can see plots of this function for values of $n$ from 1 to 3 in Fig. 2.24a; all are continuous across $t = 0$.

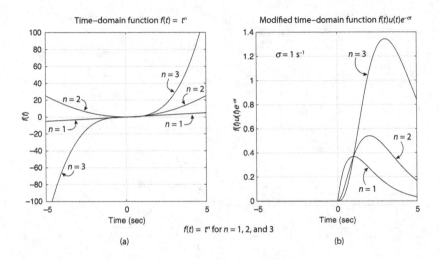

Figure 2.24 Power of $t$ time-domain function (a) $f(t)$ and (b) modified $f(t)$.

For any positive value $n$, the Fourier-transform integral of $t^n e^{-i\omega t}$ over all time does not converge, as described in Section 2.1 for $n = 0$ (constant time-domain function). But even for higher values of $n$, these integrals can be made to converge by multiplying $f(t)$ by the unit-step function $u(t)$ and the real exponential $e^{-\sigma t}$ with $\sigma > 0$.

The modified functions $f(t)u(t)e^{-\sigma t}$ for $n = 1, 2$, and 3 are shown in Fig. 2.24b. As you can see, the shapes of these functions are similar, but the steeper slopes for higher values of $n$ cause the peaks to shift to higher amplitudes and later times. This means the mixtures of cosine and sine basis functions that make up the modified functions $f(t)u(t)e^{-\sigma t}$ are similar but not identical for different values of $n$, as you will see below.

Before getting to that, here's the mathematical derivation of the unilateral Laplace transform of the function $f(t) = t^n$:

$$F(s) = \int_{0^-}^{+\infty} f(t)e^{-st}dt = \int_0^{+\infty} t^n e^{-st}dt.$$

As in previous examples, the continuity of $f(t)$ across time $t = 0$ means that the lower limit of the second integral can be written as 0. This improper integral can be related to a known function by making a change of variables $x = st$, so

$$x = st \qquad t = \frac{x}{s} \qquad \frac{dt}{dx} = \frac{1}{s} \qquad dt = \frac{dx}{s},$$

which makes the Laplace transform look like this:

$$F(s) = \int_0^{+\infty} \left(\frac{x}{s}\right)^n e^{-s\left(\frac{x}{s}\right)} \frac{dx}{s} = \left(\frac{1}{s^{(n+1)}}\right) \int_0^{+\infty} x^n e^{-x} dx. \qquad (2.18)$$

The second integral in this equation is a form of the gamma function, defined as

$$\Gamma(n) \equiv \int_0^{+\infty} x^{n+1} e^{-x} dx$$

or

$$\Gamma(n-1) = \int_0^{+\infty} x^n e^{-x} dx. \qquad (2.19)$$

Inserting this into Eq. 2.18 gives

$$F(s) = \left(\frac{1}{s^{(n+1)}}\right) \int_0^{+\infty} x^n e^{-x} dx = \frac{1}{s^{(n+1)}} \Gamma(n-1) \qquad (2.20)$$

for any value of $s$ greater than zero.

An additional simplification can be achieved using the relation between the gamma function of $(n-1)$ and $n$ factorial (written as $n!$):

$$\Gamma(n-1) = n!, \qquad (2.21)$$

which holds for any integer value of $n$ equal to or greater than zero (note that $0!$ is defined as one). Thus

$$F(s) = \frac{1}{s^{(n+1)}} \Gamma(n-1) = \frac{n!}{s^{(n+1)}}. \qquad (2.22)$$

So for $f(t) = t^n$, the integral of $f(t)e^{-st}$ over the range $t = 0$ to $t = \infty$ converges to a value of $\frac{n!}{s^{(n+1)}}$ as long as $s > 0$:

$$F(s) = \frac{n!}{s^{(n+1)}} \qquad \text{ROC: } s > 0. \qquad (2.23)$$

For $n = 0$, this reduces to the expression shown in Eq. 2.4 for a constant time-domain function $f(t) = c$ with $c = 1$:

$$F(s) = \frac{n!}{s^{(n+1)}} = \frac{0!}{s^{(0+1)}} = \frac{1}{s},$$

in which the relation $0! = 1$ has been used.

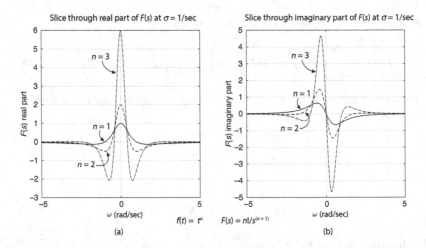

Figure 2.25  (a) Real and (b) imaginary parts of a slice through $F(s) = n!/s^{(n+1)}$ at $\sigma = 1/\text{sec}$.

Here are the equations for the real and imaginary parts of $F(s)$ for $n = 1$. For the real part

$$\text{Re}\left[F(s)\right] = \text{Re}\left[\frac{n!}{s^{(n+1)}}\right] = \text{Re}\left[\frac{1!}{s^{(1+1)}}\right]$$

$$= \text{Re}\left[\frac{1}{s^2}\right] = \text{Re}\left[\frac{1}{(\sigma + i\omega)^2}\right],$$

which can be rationalized to give

$$\text{Re}\left[F(s)\right] = \frac{\sigma^2 - \omega^2}{(\sigma^2 + \omega^2)^2} \tag{2.24}$$

(if you need some help seeing why this is true, take a look at the chapter-end problems and online solutions).

For the imaginary part

$$\text{Im}\left[F(s)\right] = \text{Im}\left[\frac{1}{s^2}\right] = \text{Im}\left[\frac{1}{(\sigma + i\omega)^2}\right],$$

which rationalizes to

$$\text{Im}\left[F(s)\right] = -\frac{2\omega\sigma}{(\sigma^2 + \omega^2)^2}. \tag{2.25}$$

The behavior of the real and imaginary parts of $F(s)$ is shown graphically in the slices through $F(s)$ shown in Fig. 2.25; these slices are taken at $\sigma = 1/\text{sec}$ for $n = 1$, 2, and 3. For all values of $n$, the real parts are symmetric with a single positive peak at $\omega = 0$, indicative of a large positive DC component, with smaller negative lobes on both sides of the peak. As you'll see in the discussion of the inverse Laplace transform later in this section, those negative lobes add negative cosine basis functions into the mix that forms the modified time-domain function $f(t)u(t)e^{-\sigma t}$, reducing the integrated result for times $t < 0$ when combined with the positive cosine components and the sine components contributed by the imaginary part of $F(s)$.

Likewise, at all values of $n$, the imaginary parts of $F(s)$ are antisymmetric, with large positive peaks at negative angular frequencies ($\omega < 0$) and large negative peaks at positive angular frequencies ($\omega > 0$). For negative frequencies, these positive imaginary peaks (which contain a multiplicative factor of $+i$) are multiplied by $i \sin(-\omega t) = -i \sin(\omega t)$, adding positive sine basis functions to the mix in the inverse Laplace transform. For positive frequencies, the negative imaginary peaks (which contain a multiplicative factor of $-i$) are multiplied by $i \sin(\omega t)$, also adding positive sine basis functions. As described in the discussion of the inverse Laplace transform below, these low-frequency positive sine components produce the required slope and peak position for the modified time-domain function $f(t)u(t)e^{-\sigma t}$ shown in Fig. 2.24b. These components also contribute to the cancellation of the positive cosine components for times $t < 0$, as needed for the modified time-domain function $f(t)u(t)e^{-\sigma t}$ to have zero amplitude for all negative times.

To see how this works, consider the three low-frequency components of $F(s)e^{st}$ at angular frequencies of $\omega = \pm 0.025$ rad/sec, $\pm 0.065$ rad/sec, and $\pm 0.15$ rad/sec shown in Fig. 2.26 for $n = 1$ and $\sigma = 0.1$/sec. At the lower two of these three angular frequencies, the components are combinations of positive cosine functions from the real part of $F(s)$ and positive sine functions from the imaginary part of $F(s)$. Since cosine functions have a positive peak at time $t = 0$ and sine functions have a positive peak one quarter-cycle to the right of that time, the combination of these functions produces a time-domain function with a peak just to the right of $t = 0$, with the exact location determined by the relative amounts of cosine and sine functions (that is, the ratio of the real part of $F(s)$ to the imaginary part). Wherever that peak occurs, you can expect these components to have increasing amplitude for some fraction of a cycle as time increases from $t = 0$, and you can see that happening for each of the three components over the time window shown in the middle portion of this figure (which is considerably shorter than one cycle for these low frequencies). That behavior is helpful in synthesizing the time-domain function $f(t) = t^n$, which is a linearly increasing function in this case since $n = 1$. At the highest of these three frequencies ($\omega = \pm 0.15$ rad/sec) the cosine component is negative and quite small relative to the sine

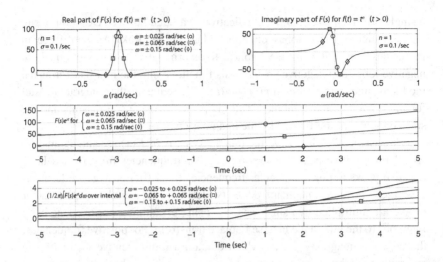

Figure 2.26 Three low-frequency components of $F(s)$ for $f(t) = t^n$.

component, as you can see in the top portion of this figure, and that's why the amplitude of this component is small and negative for times less than $t = 0$, as shown in the middle portion of this figure. The integrated results over these narrow frequency ranges have smaller amplitudes and slopes for $t < 0$ than for $t > 0$, as shown in the bottom portion of this figure, but it's clear that additional frequency components are needed to drive the integrated results closer to the function $f(t)$ with zero amplitude for $t < 0$ and constant slope of unity for $t > 0$.

You can see the effect of mixing in additional frequency components in Fig. 2.27, which shows the medium-frequency components $\omega = \pm 0.3$ rad/sec, $\pm 0.65$ rad/sec, and $\pm 2.0$ rad/sec. As shown in the middle portion of this figure, the amplitudes of these components are smaller than those of the low-frequency components shown in the previous figure, but integrating over these wider frequency ranges goes a long way to pushing the slope of the integrated result toward zero for $t < 0$ and one for $t > 0$. This is illustrated in the bottom portion of the figure, in which the result of integrating $F(s)e^{st}$ from $-2$ rad/sec to $+2$ rad/sec has the small slope and amplitude for times $t < 0$ as well as the required linearly increasing amplitude for time $t > 0$. Closer inspection reveals some departure from $f(t) = t$ as well as amplitude ripples, and both of those effects can be reduced by mixing in higher-frequency components.

The components at frequencies of $\omega = \pm 3$ rad/sec, $\pm 5$ rad/sec, and $\pm 10$ rad/sec are shown in Fig. 2.28. At these frequencies, the negative cosine

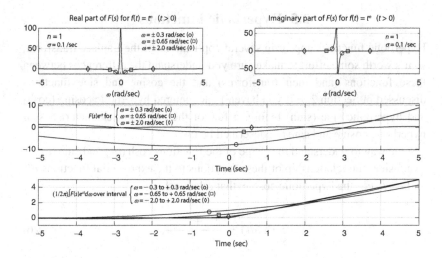

Figure 2.27 Three mid-frequency components of $F(s)$ for $f(t) = t^n$.

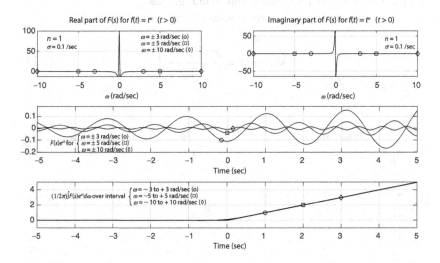

Figure 2.28 Three high-frequency components of $F(s)$ for $f(t) = t^n$.

components dominate, as you can see in the middle portion of the figure. Adding in these increasingly small components makes the integrated result almost indistinguishable from $f(t) = 0$ for $t < 0$ and $f(t) = t$ for $t > 0$ at the scale shown in the bottom portion of this figure.

## 2.5 Hyperbolic Functions

Hyperbolic functions appear in several applications of the Laplace transform, so it's worth some effort to make sure you understand the differences between these functions (and their transforms) and the cosine and sine functions discussed in Section 2.3. If you haven't encountered hyperbolic cosine (cosh) and hyperbolic sine (sinh) before, a few of the important characteristics of hyperbolic cosine and sine are described below.

One way to understand the shape of the functions $\cosh(at)$ and $\sinh(at)$ is to consider the relationship of these functions to the exponential functions $e^{at}$ and $e^{-at}$. For the hyperbolic cosine, that relationship is

$$\cosh(at) = \frac{e^{at} + e^{-at}}{2}, \tag{2.26}$$

which says that the hyperbolic cosine is the average of the exponential functions $e^{at}$ and $e^{-at}$. A plot of these two functions and the cosh function for positive time and $a = 1/\sec$ is shown in Fig. 2.29a.

For the hyperbolic sine, the relationship to exponential functions is

$$\sinh(at) = \frac{e^{at} - e^{-at}}{2}, \tag{2.27}$$

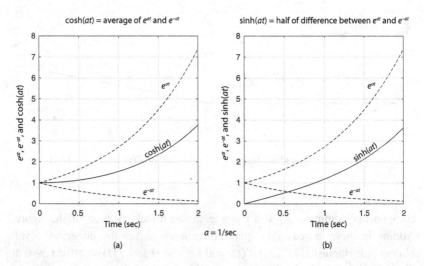

Figure 2.29 Relationship of (a) hyperbolic cosine and (b) hyperbolic sine to the exponential functions $e^{at}$ and $e^{-at}$.

so the hyperbolic sine is half of the difference between the exponential functions $e^{at}$ and $e^{-at}$, as shown for positive time and $a = 1/\text{sec}$ in Fig. 2.29b.

Although the shape of the cosh function and the sinh function are similar and both are continuous across time $t = 0$, note that the cosh function has a value of unity and slope of zero at time $t = 0$, while the sinh function has a value of zero and slope of $a$ (unity in this example) at $t = 0$. As you'll see below, those values have an impact on the Laplace transforms of these hyperbolic functions.

If you are wondering about the meaning of the parameter $a$ in the functions $\cosh(at)$ and $\sinh(at)$, note that the dimensions of $a$ must be inverse time (SI units of 1/sec) in order to render the exponent $(at)$ dimensionless. Thus $a$, as it was in Eq. 2.8 ($f(t) = e^{at}$) of Section 2.2, is a form of frequency, telling you how quickly the functions $\cosh(at)$ and $\sinh(at)$ change with time.

For positive, real values of $a$, the difference between the $\cosh(at)$ function and the $\sinh(at)$ function becomes smaller as time increases, since in that case the amplitude of the term $e^{-at}$ in Eqs. 2.26 and 2.27 becomes small relative to $e^{at}$. That means that both $\cosh(at)$ and $\sinh(at)$ approach the value of $e^{at}/2$ at large values of $t$; it doesn't make much difference if you add or subtract the $e^{-at}$ term. So large values of $a$ mean greater curvature, steeper slope, and smaller difference between the time-domain functions $\cosh(at)$ and $\sinh(at)$.

### Hyperbolic Cosine Functions

The process of finding the unilateral Laplace transform of the hyperbolic cosine function

$$f(t) = \cosh(at) \tag{2.28}$$

will look familiar to you if you've worked through Section 2.3 above. As in that case, start by writing

$$F(s) = \int_{0^-}^{+\infty} f(t)e^{-st}dt = \int_{0}^{+\infty} \cosh(at)e^{-st}dt. \tag{2.29}$$

As in previous examples, the continuity of the time-domain function $f(t)$ across time $t = 0$ means that the lower limit of integration can be set to 0 rather than $0^-$.

You can see a plot of the time-domain function $f(t) = \cosh(at)$ for $a = 1/\text{sec}$ in Fig. 2.30a and the modified function $f(t)u(t)e^{-\sigma t}$ for $\sigma = 1/\text{sec}$ in Fig. 2.30b.

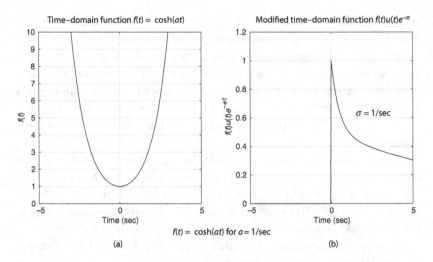

Figure 2.30 (a) Hyperbolic cosine time-domain function $f(t)$ and (b) modified $f(t)$.

Inserting the expression for $\cosh at$ from Eq. 2.26 into Eq. 2.29 gives

$$F(s) = \int_0^{+\infty} \left[ \frac{e^{at} + e^{-at}}{2} \right] e^{-st} dt = \int_0^{+\infty} \left[ \frac{e^{(a-s)t} + e^{(-a-s)t}}{2} \right] dt$$

$$= \frac{1}{2} \left[ \int_0^{+\infty} e^{-(s-a)t} dt + \int_0^{+\infty} e^{-(s+a)t} dt \right].$$

As in previous examples, these improper integrals can be evaluated by calling the upper limit of integration $\tau$ and taking the limit as $\tau$ goes to infinity:

$$F(s) = \lim_{\tau \to \infty} \frac{1}{2} \left[ \int_0^{\tau} e^{-(s-a)t} dt + \int_0^{\tau} e^{-(s+a)t} dt \right]$$

$$= \lim_{\tau \to \infty} \frac{1}{2} \left[ \frac{1}{-(s-a)} e^{-(s-a)t} \right]_0^{\tau} + \lim_{\tau \to \infty} \frac{1}{2} \left[ \frac{1}{-(s+a)} e^{-(s+a)t} \right]_0^{\tau}$$

$$= \lim_{\tau \to \infty} \frac{1}{2} \left[ \frac{1}{-(s-a)} e^{-(s-a)\tau} - \frac{1}{-(s-a)} e^{-(s-a)0} \right]$$

$$+ \lim_{\tau \to \infty} \frac{1}{2} \left[ \frac{1}{-(s+a)} e^{-(s+a)\tau} - \frac{1}{-(s+a)} e^{-(s+a)0} \right].$$

Inside the first set of large square brackets, the exponential term in which $\tau$ appears will be zero as long as $s - a > 0$ (so $s > a$), and in the second pair of large square brackets, the exponential term in which $\tau$ appears will be zero as long as $s + a > 0$ (so $s > -a$). To cover both of these cases for both positive and negative values of $a$, $s$ must be greater than $|a|$. If that's true, then both terms involving $\tau$ go to zero as $\tau \to \infty$, leaving

$$F(s) = \frac{1}{2}\left[0 - \frac{1}{-(s-a)}e^{-(s-a)0}\right] + \frac{1}{2}\left[0 - \frac{1}{-(s+a)}e^{-(s+a)0}\right]$$

$$= \frac{1}{2}\left[\frac{1}{(s-a)} + \frac{1}{(s+a)}\right].$$

Combining these two terms gives

$$F(s) = \frac{1}{2}\left[\frac{s+a}{(s-a)(s+a)} + \frac{s-a}{(s+a)(s-a)}\right]$$

$$= \frac{1}{2}\left[\frac{s+a+s-a}{s^2-a^2}\right] = \frac{1}{2}\left[\frac{2s}{s^2-a^2}\right] = \frac{s}{s^2-a^2}.$$

So in this case the integral of $f(t)e^{-st}$ over the range $t = 0$ to $t = \infty$ converges to a value of $\frac{s}{s^2-a^2}$ as long as $s > |a|$. Thus the unilateral Laplace transform of the hyperbolic cosine function $f(t) = \cosh{(at)}$ is

$$F(s) = \frac{s}{s^2-a^2} \qquad \text{ROC: } s > |a|. \qquad (2.30)$$

The real and imaginary parts of $F(s)$ for $a = 1/\text{sec}$ and $\sigma = 1.1/\text{sec}$ are shown in Fig. 2.31a and 2.31b. As expected for the transform of a time-domain signal with a large DC component, the frequency components with the highest amplitude are the low-frequency cosine basis functions represented by the real part of $F(s)$ near $\omega = 0$.

But sine basis functions are also needed to produce the slope of the modified time-domain function $f(t)u(t)e^{-\sigma t}$, and you can see the amplitudes of those components in the slice through the imaginary part of $F(s)$ shown in Fig. 2.31b. As described in previous examples, the anti-symmetric nature of the imaginary part of $F(s)$ about $\omega = 0$ means that the sine components from both sides of the spectrum have the same sign, since $\sin{(\omega t)}$ is also an odd function. In this case, the positive imaginary values of $F(s)$ for negative angular frequencies ($\omega < 0$) multiplied by $i \sin{(-\omega t)}$ produce positive sine components. When synthesizing the time-domain function $f(t)$ using the inverse Laplace transform, these positive sine components provide the required slope for positive time and combine with the positive cosine components to reduce the amplitude of the integrated result for negative time.

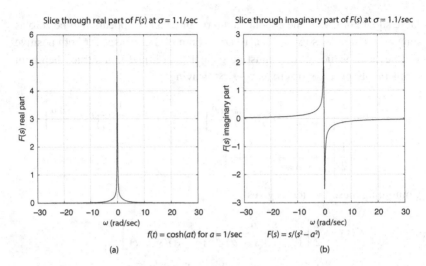

(a)

(b)

Figure 2.31 (a) Real and (b) imaginary parts of a slice through $F(s) = s/(s^2 + a^2)$ at $\sigma = 1.1/$sec and $a = 1/$sec.

You can see how the cosine and sine basis functions weighted by the real and imaginary parts of $F(s)$ combine to product the time-domain function $f(t) = \cosh(at)$ for $t > 0$ by considering the behavior over time of several of the low-, medium-, and high-frequency components.

As in the case of the examples of constant and exponential functions discussed earlier in this chapter, the low-frequency components of the inverse Laplace transform establish the small-amplitude baseline for negative time and increasing amplitude for positive time. You can see three of those low-frequency components for $F(s) = s/(s^2 - a^2)$ in Fig. 2.32; those frequencies are $\omega = \pm 0.01$ rad/sec, $\pm 0.05$ rad/sec, and $\pm 0.1$ rad/sec. At each of these frequencies, the real part of $F(s)$ contributes positive cosine functions and the imaginary part of $F(s)$ contributes positive sine functions to the inverse Laplace transform. Those functions mix to provide components that have small but increasing amplitude over the $\pm 5$-second time window shown in the middle portion of this figure. Note that the slope of the integrated results is small for negative time and increasing for positive time; at the highest of these three frequencies ($\omega = \pm 0.1$ rad/sec), the slope is greater than the slope of the time-domain function $f(t) = \cosh(at)$. And, as in previous examples, these slowly changing components cannot produce the required step in amplitude at time $t = 0$, so higher-frequency components are needed to do that job. The next

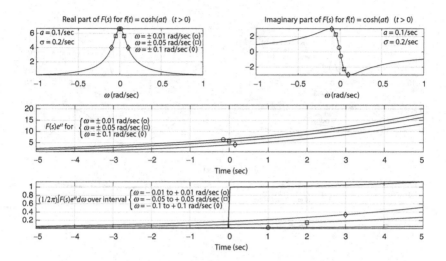

Figure 2.32  Three low-frequency components of $F(s)$ for $f(t) = \cosh(at)$.

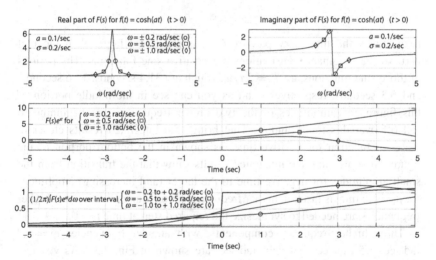

Figure 2.33  Three medium-frequency components of $F(s)$ for $f(t) = \cosh(at)$.

two plots show the effect of including higher-frequency components in the process of synthesizing $f(t)$.

Three medium-frequency components with angular frequencies of $\omega = \pm 0.2$ rad/sec, $\pm 0.5$ rad/sec, and $\pm 1.0$ rad/sec are shown in Fig. 2.33. At these

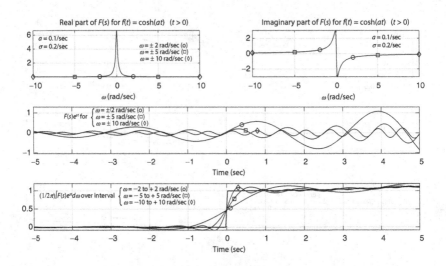

Figure 2.34  Three high-frequency components of $F(s)$ for $f(t) = \cosh(at)$.

frequencies, the imaginary part of $F(s)$ is somewhat larger than the real part, so the dominant contributions are positive sine functions. The periods of those sine functions are $T = 2\pi/\omega$, which are 31.4 seconds, 12.6 seconds, and 6.3 seconds, respectively, and as you can see in the middle portion of this figure, the amplitudes of the two higher frequencies are beginning to roll off over this time window. That brings the integrated results closer to the cosh function for positive time, as shown in the bottom portion of the figure. Note also that the integrated results show that the transition from the small amplitudes for negative time to the larger (and increasing) amplitudes for positive time is faster at higher frequencies; this is an indication that higher frequencies are needed to synthesize the required step at time $t = 0$.

Three higher-frequency components, with angular frequencies $\omega = \pm 2$ rad/sec, $\pm 5$ rad/sec, and $\pm 10$ rad/sec, are shown in Fig. 2.34. As you can see in the middle portion of this figure, each of the positive sine functions at these frequencies has a relatively steep positive slope at time $t = 0$, so including these components in the integral of the inverse Laplace transform contributes to the amplitude step of $f(t)$ as it transitions from zero amplitude for $t < 0$ to $\cosh(at)$ for $t > 0$. Including these higher-frequency components in the integral of the inverse Laplace transform also has the effect of smoothing out the ripples for both negative and positive time, but the Gibbs ripples near

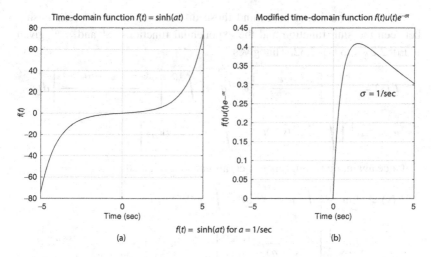

Figure 2.35 (a) Hyperbolic sine time-domain function $f(t)$ and (b) modified $f(t)$.

the amplitude jump at time $t = 0$ remain, as discussed in the constant-time example at the beginning of this chapter.

## Hyperbolic Sine Functions

As shown above for the hyperbolic cosine function, the process of finding the Laplace transform of the hyperbolic sine time-domain function

$$f(t) = \sinh(at) \qquad (2.31)$$

begins by writing the Laplace transform as

$$F(s) = \int_{0^-}^{+\infty} f(t)e^{-st}dt = \int_0^{+\infty} \sinh(at)e^{-st}dt. \qquad (2.32)$$

The time-domain function $f(t) = \sinh(at)$ for $a = 1/\text{sec}$ is shown in Fig. 2.35a and the modified function $f(t)u(t)e^{-\sigma t}$ for $\sigma = 1/\text{sec}$ in Fig. 2.35b. Although the modified function looks somewhat similar to the modified function for the hyperbolic cosine function shown in Fig. 2.30, a careful look reveals several differences that lead to differences in the Laplace transform of the time-domain sinh function from that of the cosh function discussed above.

To see the mathematics behind those differences, insert the relationship between the sinh function and the exponential functions $e^{at}$ and $e^{-at}$ given in Eq. 2.27 into Eq. 2.32. That gives

$$F(s) = \int_0^{+\infty} \left[ \frac{e^{at} - e^{-at}}{2} \right] e^{-st} dt = \int_0^{+\infty} \left[ \frac{e^{(a-s)t} - e^{(-a-s)t}}{2} \right] dt$$

$$= \frac{1}{2} \left[ \int_0^{+\infty} e^{-(s-a)t} dt - \int_0^{+\infty} e^{-(s+a)t} dt \right].$$

Once again, these improper integrals can be evaluated as:

$$F(s) = \lim_{\tau \to \infty} \frac{1}{2} \left[ \int_0^{\tau} e^{-(s-a)t} dt - \int_0^{\tau} e^{-(s+a)t} dt \right]$$

$$= \lim_{\tau \to \infty} \frac{1}{2} \left[ \frac{1}{-(s-a)} e^{-(s-a)t} \right] \Big|_0^{\tau} - \lim_{\tau \to \infty} \frac{1}{2} \left[ \frac{1}{-(s+a)} e^{-(s+a)t} \right] \Big|_0^{\tau}$$

$$= \lim_{\tau \to \infty} \frac{1}{2} \left[ \frac{1}{-(s-a)} e^{-(s-a)\tau} - \frac{1}{-(s-a)} e^{-(s-a)0} \right]$$

$$- \lim_{\tau \to \infty} \frac{1}{2} \left[ \frac{1}{-(s+a)} e^{-(s+a)\tau} - \frac{1}{-(s+a)} e^{-(s+a)0} \right].$$

As in the cosh case, if $s$ is greater than $|a|$, both terms involving $\tau$ go to zero as $\tau \to \infty$, leaving

$$F(s) = \frac{1}{2} \left[ 0 - \frac{1}{-(s-a)} e^{-(s-a)0} \right] - \frac{1}{2} \left[ 0 - \frac{1}{-(s+a)} e^{-(s+a)0} \right]$$

$$= \frac{1}{2} \left[ \frac{1}{(s-a)} - \frac{1}{(s+a)} \right].$$

In this case combining these two terms gives

$$F(s) = \frac{1}{2} \left[ \frac{s+a}{(s-a)(s+a)} - \frac{s-a}{(s+a)(s-a)} \right]$$

$$= \frac{1}{2} \left[ \frac{s+a-s+a}{s^2 - a^2} \right] = \frac{1}{2} \left[ \frac{2a}{s^2 - a^2} \right] = \frac{a}{s^2 - a^2}.$$

So the integral of $f(t)e^{-st}$ over the range $t = 0$ to $t = \infty$ converges to a value of $a/(s^2 - a^2)$ as long as $s > |a|$. Thus the unilateral Laplace transform of the hyperbolic sine function $f(t) = \sinh at$ is

$$F(s) = \frac{a}{s^2 - a^2} \qquad \text{ROC: } s > |a|. \qquad (2.33)$$

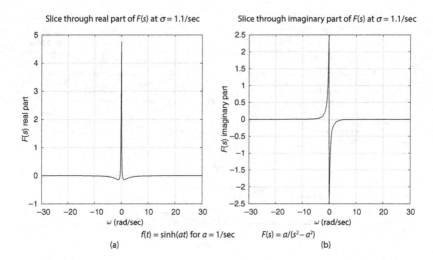

Figure 2.36 (a) Real and (b) imaginary parts of slice through $F(s)$ at $\sigma = 1.1/\text{sec}$ and $a = 1/\text{sec}$.

There are obvious similarities between $F(s)$ for the sinh function and the cosh function, as you can see in the plots of the real and imaginary parts of $F(s)$ shown in Fig. 2.36. As was the case for the Laplace transform of the time-domain cosh function, the highest-amplitude frequency components of the sinh function are low-frequency cosine functions, because like the modified cosh function, the modified sinh function has a large DC component (as shown in Fig. 2.35b above). And like the cosh transform, the sinh transform has smaller but nonzero sine components, which add to the mix in the inverse Laplace transform to produce the proper slope and to combine with the positive cosine components to reduce the amplitude of the integrated result for negative times.

But a careful comparison of $F(s)$ for sinh and cosh functions reveals important differences, as well. In Fig. 2.36a, notice the small negative-amplitude lobes on both sides of the peak at $\omega = 0$. As you'll see in the discussion of the inverse Laplace transform below, the cosine basis functions with amplitudes given by those negative lobes contribute to the cancellation of the positive cosine components not just at negative times, but also at time $t = 0$. As mentioned above, one difference between the time-domain cosh and sinh functions is that the cosh function has a value of unity at time $t = 0$, while the sinh function has a value of zero at that time. And since all sine functions have a value of zero at $t = 0$, the cancellation of positive cosine contributions can be done only by the addition of negative cosine basis functions.

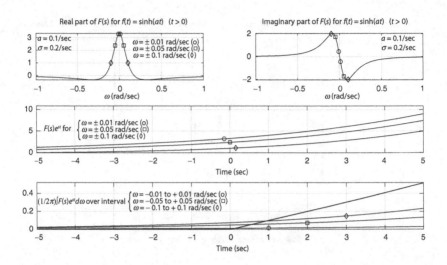

Figure 2.37  Three low-frequency components of $F(s)$ for $f(t) = \sinh(at)$.

The other substantial difference in $F(s)$ for cosh and sinh functions can be seen by comparing the slice through the imaginary part of $F(s)$ shown in Fig. 2.36b for the sinh function with the equivalent slice for the cosh function shown in Fig. 2.31b. As you can see, the spread of frequency components is significantly smaller in the sinh case (that is, the amplitude of the components rolls off more quickly with frequency in this case). This makes sense when you note that the modified time-domain function $f(t)u(t)e^{\sigma t}$ is considerably narrower for cosh than for sinh, and narrower time-domain functions require wider frequency-domain spectra, since higher frequency components are needed to produce steeper slopes in time.

As you might expect from the similarity of the $s$-domain functions $F(s)$ near $\omega = 0$ for cosh and sinh time-domain functions, the low-frequency components shown in Fig. 2.37 are similar to those of the cosh function, with the real part of $F(s)$ contributing positive cosine functions and the imaginary part of $F(s)$ contributing positive sine functions. In both cases, the resulting amplitude of each of these components is small for $t < 0$ and increasing for $t > 0$. But note that the sinh function has no step at time $t = 0$, and the amplitude of the integrated results is smaller than $\sinh(at)$ for positive times.

That means that additional frequency components are needed to increase the amplitude of the integrated results at positive time, but unlike the cosh case,

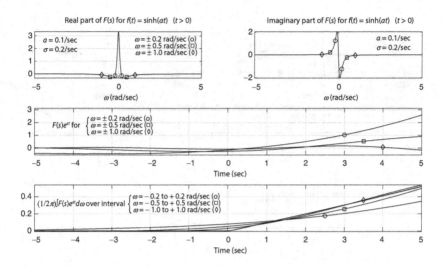

Figure 2.38 Three medium-frequency components of $F(s)$ for $f(t) = \sinh(at)$.

those components must maintain the continuous (and small) amplitude across time $t = 0$.

To see how that works, take a look at the three medium-frequency components with $\omega = \pm 0.2$ rad/sec, $\pm 0.5$ rad/sec, and $\pm 1.0$ rad/sec shown in the middle portion of Fig. 2.38. These medium-frequency components contribute positive sine functions from the imaginary part of $F(s)$, but some of the cosine functions contributed by the real part of $F(s)$ are negative. As you can see in the integrated results shown in the bottom portion of this figure, those negative cosine functions help prevent a step from occurring at time $t = 0$.

The result of increasing the range of integration to $\omega = \pm 2$ rad/sec, $\pm 5$ rad/sec, and $\pm 10$ rad/sec is shown in the Fig. 2.39. Integrating over even the most narrow of these frequency ranges produces a function that closely resembles $f(t) = 0$ for $t < 0$ and $f(t) = \sinh(at)$ for $t > 0$, but, as expected, higher-frequency components contribute to the sharpness of the change in slope at time $t = 0$ and reduce the amplitude ripples for both negative and positive time.

Since the integrated results are virtually indistinguishable from $f(t) = \sinh(at)$ at the scale shown in Fig. 2.39, you can get a better idea of the contributions of high-frequency components by looking at the zoomed-in plot in Fig. 2.40. In this plot, you can see that the addition of higher frequencies

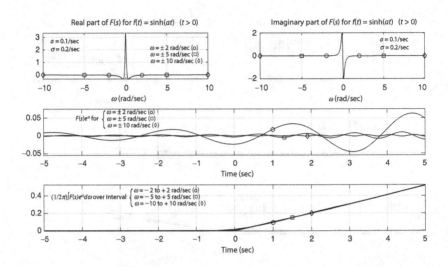

Figure 2.39 Three high-frequency components of $F(s)$ for $f(t) = \sinh(at)$.

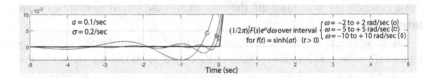

Figure 2.40 Closer view of the integrated results of $F(s)e^{st}$ over the frequency ranges of $\pm 2$, $\pm 5$, and $\pm 1$ rad/sec for $f(t) = \sinh(at)$.

sharpen the transition in amplitude from zero for $t < 0$ to $\sinh(at)$ for $t > 0$, while also reducing the amplitude and period of the ripples in the synthesized result.

With the Laplace transforms of these constant, exponential, sinusoidal, $t^n$, and hyperbolic functions in hand, you should be ready to see and understand some characteristics of the Laplace transform. Those characteristics are the subject of the next chapter, which you should find easy going if you take the time to work through each of the problems in the following section – and remember, if you need help, full interactive solutions are available on this book's website.

## 2.6 Problems

1. The Fourier and Laplace transforms involve the integral of the product of
   the complex-exponential basis functions and the time-domain function
   $f(t)$; the result depends on the even or odd nature of those functions.
   Show that

   (a) Multiplying two even functions produces an even function.
   (b) Multiplying two odd functions produces an even function.
   (c) Multiplying an even and an odd function produces an odd function.
   (d) Any function $f(t)$ may be decomposed into the sum
      $f(t) = f_{even}(t) + f_{odd}(t)$ of an even and an odd function defined by

   $$f_{even}(t) = \frac{f(t) + f(-t)}{2} \qquad f_{odd}(t) = \frac{f(t) - f(-t)}{2}.$$

2. The unilateral Laplace transform of the constant time-domain function
   $f(t) = c$ is discussed in Section 2.1. Use the same approach to find the
   $s$-domain function $F(s)$ and the region of convergence if the
   time-domain function $f(t)$ is limited in time, specifically if $f(t) = 2$ for
   $0 < t < 1$ and $f(t) = 0$ for $t > 1$.

3. Find the real and imaginary parts of the unilateral Laplace transform
   $F(s)$ of the exponential time-domain function $f(t) = e^{at}$ discussed in
   Section 2.2 and specify whether each is even or odd.

4. Use the approach of Section 2.2 and the results of the previous problem
   to find the real and imaginary parts of $F(s)$ and the ROC for the
   time-domain function $f(t) = 5e^{-3t}$.

5. Sketch the scaled time-domain cosine function $f(t) = \frac{\cos(2t)}{4}$ and use the
   definition of the unilateral Laplace transform (Eq. 1.2) to find $F(s)$ for
   this function.

6. Use the definition of the unilateral Laplace transform to find $F(s)$ for the
   time-offset sine function $f(t) = \sin(t - \frac{\pi}{4})$ for $t \geq \frac{\pi}{4}$ and $f(t) = 0$ for
   $t < \frac{\pi}{4}$.

7. Use the definition of the unilateral Laplace transform to find $F(s)$ for
   $f(t) = t$, then compare your result to Eq. 2.23 for $n = 1$. Also show that
   the expressions for the real and imaginary parts of $F(s)$ given in
   Eqs. 2.24 and 2.25 are correct.

8. Use the definition of the unilateral Laplace transform and the approach of
   Section 2.4 to find $F(s)$ for the time-domain function $f(t) = (t - 2)^2$.

9. Use the definition of the unilateral Laplace transform and the approach of
   Section 2.5 to find $F(s)$ for the time-domain function $f(t) = \cosh(\frac{t}{2})$.

10. Find the unilateral Laplace transform of the time-domain function
    $f(t) = 6\cosh^2(-4t) - 3\sinh(2t)$.

# 3

# Properties of the Laplace Transform

The value of knowing the Laplace transforms of the basic functions described in the previous chapter is greatly enhanced by certain properties of the Laplace transform. That is because these properties allow you to determine the transform of much more complicated time-domain functions by combining and modifying the transforms of simple functions such as those discussed in Chapter 2.

In this chapter, you will find descriptions of the most important characteristics of the Laplace transform. Those characteristics include linearity (Section 3.1), shifting (Section 3.2), scaling (Section 3.3), time-domain differentiation (Section 3.4), integration (Section 3.5), multiplication and division by $t$ (Section 3.6), periodic functions (Section 3.7), convolution (Section 3.8), and the initial- and final-value theorems (Section 3.9). These properties apply to the Laplace transforms of time-domain functions $f(t)$ that are piecewise continuous and of exponential order. The descriptions in this chapter include explanations of the mathematical foundations of each property, but they also provide an overview of the reasoning behind each characteristic – that is, why it "makes sense" that the Laplace transform behaves as it does.

As in every chapter, the final section of this chapter (Section 3.10) contains a set of problems that you can use to check your understanding of the concepts and mathematical techniques presented in this chapter.

## 3.1 Linearity

A very useful characteristic of the Laplace transform is linearity, which means that transforming a time-domain function $f(t)$ to a generalized-frequency domain function $F(s)$ using the Laplace transform is a linear process. A linear process encompasses two properties: homogeneity and additivity.

The homogeneity of the Laplace transform means that if the time-domain function $f(t)$ has Laplace transform $F(s)$, then the Laplace transform of the function $cf(t)$ (in which $c$ is a constant) is simply $cF(s)$. You can see this by writing out the one-sided Laplace transform as

$$\mathcal{L}[f(t)] = \int_{0^-}^{\infty} f(t)e^{-st}dt = F(s),$$

so the Laplace transform of $cf(t)$ is

$$\mathcal{L}[cf(t)] = \int_{0^-}^{\infty} cf(t)e^{-st}dt = c\int_{0^-}^{\infty} f(t)e^{-st}dt = cF(s).$$

This can be stated concisely as

$$\mathcal{L}[cf(t)] = c\mathcal{L}[f(t)] = cF(s). \tag{3.1}$$

Intuitively, this should make sense to you, because $F(s)$ tells you how much of each cosine and sine basis function is included in the mix that makes up the modified time-domain function $f(t)u(t)e^{-\sigma t}$. So scaling the amplitude of all of the basis functions by multiplying by a constant also scales the results of mixing those functions together (in other words, if you're baking a cake and you want to make it twice as big, you should double the amount of each ingredient you put in).

The additivity of the Laplace transform means that if you have a time-domain function that you recognize as the sum of two or more other functions, the Laplace transform of that composite function is just the sum of the Laplace transforms of each of the individual functions. So for two time-domain functions $f(t)$ and $g(t)$ with Laplace transforms $F(s)$ and $G(s)$, the Laplace transform of the sum $f(t) + g(t)$ is just the sum of the Laplace transforms $F(s) + G(s)$. That is, if

$$F(s) = \int_{0^-}^{\infty} f(t)e^{-st}dt \quad \text{and} \quad G(s) = \int_{0^-}^{\infty} g(t)e^{-st}dt$$

then

$$\mathcal{L}[f(t) + g(t)] = \int_{0^-}^{\infty} [f(t) + g(t)]e^{-st}dt$$

$$= \int_{0^-}^{\infty} f(t)e^{-st}dt + \int_{0^-}^{\infty} g(t)e^{-st}dt$$

$$= F(s) + G(s).$$

In operator notation, this is

$$\mathcal{L}[f(t) + g(t)] = \mathcal{L}[f(t)] + \mathcal{L}[g(t)] = F(s) + G(s). \tag{3.2}$$

This should also seem reasonable to you, since all of the basis functions weighted by $F(s)$ add up to $f(t)u(t)e^{-\sigma t}$ and all of the basis functions weighted by $G(s)$ add up to $g(t)u(t)e^{-\sigma t}$. Hence adding both sets of weighted basis functions together produces $[f(t) + g(t)]u(t)e^{-\sigma t}$.

You can see how the linearity property of the Laplace transform can be used in several of the problems at the end of this chapter and their online solutions.

## 3.2 Time and Frequency Shifting

In some applications in which the Laplace transform is useful, you may encounter time-domain functions that are shifted to earlier or later times. Alternatively, you may see frequency-domain functions that are shifted to higher or lower frequencies. For example, the time-domain function $f(t - a)$ has the same shape as the function $f(t)$, but that shape is shifted to later times by an amount $a$ (if $a$ is positive) or to earlier times (if $a$ is negative).[1] Likewise, the frequency-domain function $F(s - a)$ has the same shape as the function $F(s)$, but that shape is shifted to higher complex frequencies ($s = \sigma + i\omega$) if $a$ is positive and lower complex frequencies if $a$ is negative.

In this section, you will see the consequences of such shifts on the other side of the Laplace transform – that is, the effect of a time shift such as $f(t - a)$ on the frequency-domain function $F(s)$ and the effect of a complex-frequency shift such as $F(s - a)$ on the time-domain function $f(t)$.

### Shifting in the $t$-Domain

To understand the effect of a time shift in $f(t)$ on the complex frequency function $F(s)$, start by writing the one-sided Laplace transforms of two time-domain functions: the first is $f(t)$, and the second is $f(t - a)$. In this case, $f(t - a)$ is a version of $f(t)$ that is shifted in time by amount $a$, and $f(t - a)$ has zero amplitude for time $t < a$. Here are those transforms:

$$\mathcal{L}[f(t)] = \int_{t=0^-}^{t=\infty} f(t)e^{-st}\,dt = F(s) \qquad (3.3)$$

---

[1] If you are uncertain why subtracting a positive constant from time $t$ shifts $f(t)$ to later times, take a look at the first part of the "Convolution" document on this book's website.

and

$$\mathcal{L}[f(t-a)] = \int_{t=0^-}^{t=\infty} f(t-a)u(t-a)e^{-st}dt, \qquad (3.4)$$

in which $u(t-a)$ represents a time-shifted version of the unit-step function that has amplitude zero for $t < a$ and one for $t \geq a$. Note that $F(s)$ is the result of the Laplace transform of the unshifted time-domain function $f(t)$.

Now make a new time variable ($\tau$) that is offset from time $t$ by amount $a$:

$$\tau = t - a \qquad \text{or} \qquad t = \tau + a.$$

Since the difference between $t$ and $\tau$ is just the offset $a$, an increment in time $t$ is the same as an increment in the offset time $\tau$, which means

$$dt = d\tau.$$

Inserting these expressions for $t$ and $dt$ into Eq. 3.4 for the Laplace transform of $f(t-a)$ gives

$$\mathcal{L}[f(t-a)] = \int_{t=0^-}^{t=\infty} f(t-a)u(t-a)e^{-st}dt$$

$$= \int_{\tau+a=0^-}^{\tau+a=\infty} f(\tau)u(\tau)e^{-s(\tau+a)}d\tau$$

$$= \int_{\tau=a^-}^{\tau=\infty} f(\tau)u(\tau)e^{-st}e^{-sa}d\tau,$$

in which $a^-$ represents the time $0^- - a$. But $f(\tau)u(\tau) = 0$ for $\tau < 0$, which means that setting the lower limit of integration to $0^-$ rather than $a^-$ won't change the result. And since the term $e^{-sa}$ doesn't depend on $\tau$, it can be brought out of the integral, giving

$$\mathcal{L}[f(t-a)] = e^{-sa} \int_{\tau=0^-}^{\tau=\infty} f(\tau)e^{-st}d\tau.$$

The integral in this expression is the same as the integral in Eq. 3.3 (the name of the time variable doesn't affect the result), so

$$\mathcal{L}[f(t-a)] = e^{-sa} \int_{t=0^-}^{t=\infty} f(t)e^{-st}dt = e^{-sa}F(s).$$

Written with the Laplace operator, this is

$$\mathcal{L}[f(t-a)] = e^{-sa}\mathcal{L}[f(t)] = e^{-sa}F(s). \qquad (3.5)$$

So a time-domain function $f(t-a)$, shifted in time by amount $a$ from the function $f(t)$ and with zero amplitude for $t < a$, has a Laplace transform that is the same as the Laplace transform $F(s)$ of the unshifted function $f(t)$ multiplied by the complex-exponential factor $e^{-sa}$.

Why should a shift in the time domain result in multiplication by an exponential in the frequency domain? To understand that, recall that the result $F(s)$ of the unilateral Laplace transform tells you the amount of each cosine and each sine basis function that make up the time-domain function $f(t)$, modified by multiplication by the unit-step function $u(t)$ and the real exponential function $e^{-\sigma t}$. So if $f(t)$ is shifted in time, the mixture of basis functions that make up the modified function $f(t-a)u(t-a)e^{-\sigma(t-a)}$ must be modified. Using $e^{sa}F(s)$ rather than $F(s)$ as the weighting factor of each sinusoidal basis function changes the phase of those sinusoids, and that has the effect of shifting them in time. And shifting the basis functions in time also shifts the function $f(t)$ that is synthesized by the combination of those basis functions.

### Shifting in the $s$-Domain

The mathematics of finding the effect of a complex-frequency shift such as $F(s-a)$ on the time-domain function $f(t)$ is quite straightforward. Start by writing the Laplace transform equation using $s-a$ in place of $s$:

$$F(s-a) = \int_{0^-}^{\infty} f(t)e^{-(s-a)t}dt = \int_{0^-}^{\infty} f(t)e^{-st}e^{at}dt$$

$$= \int_{0^-}^{\infty} [e^{at}f(t)]e^{-st}dt = \mathcal{L}[e^{at}f(t)]$$

or

$$\mathcal{L}[e^{at}f(t)] = F(s-a). \tag{3.6}$$

This means that a shift in the complex frequency $s$ by amount $-a$ in the frequency domain is equivalent to multiplication of $f(t)$ by the exponential $e^{at}$ in the time domain.

Why this is true (beyond the mathematics shown above) can be understood by recalling that the complex frequency $s$ is comprised of a real part $\sigma$ and an imaginary part $\omega$. Recall also that the real part $\sigma$ determines the rate of change of the factor $e^{-\sigma t}$ in the modified function $f(t)u(t)e^{-\sigma t}$, and it is this modified function for which the Laplace transform provides the basis-function amplitudes $F(s)$. But subtracting $a$ from $\sigma$ makes the modified time-domain function $f(t)u(t)e^{-(\sigma-a)t}$, which is the same as $[e^{at}f(t)]u(t)e^{-\sigma t}$.

So $F(s - a)$ tells you the weighting factors of the basis functions that combine to produce $e^{at} f(t)$ rather than $f(t)$.

To comprehend why this is true, it may help to think about a specific case such as the Laplace transform of the time-domain function $\sin(\omega_1 t)$. The Laplace transform $F(s)$ of this function tells you the amplitudes of the cosine and sine basis functions that add up to produce the modified function $f(t)u(t)e^{-\sigma t}$, which is a sine function that begins at time $t = 0$ and has decreasing amplitude as time increases. The function $F(s-a)$ also provides the weighting factors of sinusoidal basis functions, but this mix of basis functions adds up to $f(t)u(t)e^{-(\sigma - a)t}$, which is also a sine function starting at time $t = 0$ and decreasing over time. But in this case the rate of decrease is smaller, since $\sigma - a$ is smaller than $\sigma$.

And here is the key point: that smaller rate of decrease is the same as it would have been if you had started not with function $f(t) = \sin(\omega_1 t)$ (a sine function with constant amplitude), but with function $e^{at} f(t) = e^{at} \sin(\omega_1 t)$ (a sine function with increasing amplitude over time).

That is why the Laplace transform of $e^{at} f(t)$ produces the complex frequency function $F(s - a)$. And although it's tempting to think of $F(s - a)$ as "shifting to a lower frequency," you should remember that reducing the real part $\sigma$ of the complex frequency reduces the rate of change of the real exponential term $e^{\sigma t}$ but does not change the angular frequency $\omega$ of the oscillating term $e^{i\omega t}$.

You can apply your understanding of the time- and frequency-shifting property of the Laplace transform to several functions by working Problem 3.2 at the end of this chapter.

## 3.3 Scaling

The scaling property of the Laplace transform refers to the effect of a multiplicative factor in the argument of the time-domain factor $f(t)$. Calling that factor "$a$" makes the time-domain function $f(at)$, for which the Laplace transform look like this:

$$\mathcal{L}[f(at)] = \int_{0^-}^{\infty} f(at)e^{-st}dt. \qquad (3.7)$$

The impact of this scaling factor can be seen by making a change of variables:

$$u = at \qquad t = \frac{u}{a} \qquad \frac{du}{dt} = a \qquad dt = \frac{du}{a}$$

and inserting these expressions for $t$ and $dt$ into Eq. 3.7 gives

$$\mathcal{L}[f(at)] = \int_{0^-}^{\infty} f(at)e^{-st}dt = \int_{0^-}^{\infty} f(u)e^{-s\left(\frac{u}{a}\right)}\frac{du}{a}$$

$$= \frac{1}{a}\int_{0^-}^{\infty} f(u)e^{-\left(\frac{s}{a}\right)u}du.$$

But this integral is the Laplace transform of $f(u)$ with $s/a$ in place of $s$, so

$$\int_{0^-}^{\infty} f(u)e^{-\left(\frac{s}{a}\right)u}du = F\left(\frac{s}{a}\right)$$

and

$$\mathcal{L}[f(at)] = \frac{1}{a}\int_{0^-}^{\infty} f(u)e^{-\left(\frac{s}{a}\right)u}du = \frac{1}{a}F\left(\frac{s}{a}\right). \tag{3.8}$$

This means that multiplying the argument of the time-domain function by $a$ has two effects on the complex frequency-domain function $F(s)$: the argument $s$ is divided by $a$, and the amplitude of $F(s)$ is also divided by the same factor.

To understand why that is true, remember that multiplying the argument of a time-domain function $f(t)$ by a constant greater than one compresses the plot of the function along the horizontal axis, since the argument $at$ reaches a given value sooner (that is, at smaller values of time) than the argument $t$. So, as shown in Fig. 3.1a, the plot of a rectangular-pulse function $f(t)$ extending

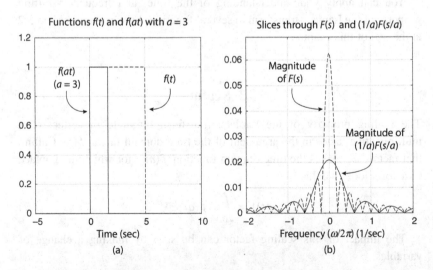

Figure 3.1 Effect of time scaling on the rectangular function $f(t) = 1$ for $0 < t < 5$ sec in (a) the time domain and (b) the frequency domain.

from $t = 0$ sec to $t = 5$ sec will extend only from $t = 0$ sec to $t = 1.67$ sec if $f(at)$ is plotted with $t$ on the horizontal axis and $a = 3$.

Recall also that compressing a time-domain function horizontally means that changes in the value of the function happen more quickly (that is, a given change occurs over a shorter period of time). That means that larger high-frequency components are needed to make up that faster-changing function. And just as multiplying a function's argument by a constant $a > 1$ compresses the plot of that function, dividing a function's argument by that constant "stretches out" that function along the horizontal axis. In the frequency domain, that means that $F(s/a)$ is wider in frequency than $F(s)$, as shown in Fig. 3.1b, exactly as needed for the compressed time-domain function.

What about the factor of $1/a$ outside $F(s/a)$? Remember that the value $f(0)$ of the time-domain function at time $t = 0$ is proportional to the area under the curve of the frequency-domain function, since the factor $e^{i\omega t}$ in the inverse transform integral is unity when $t = 0$:

$$f(t) = \frac{1}{2\pi} \int_{\sigma-i\infty}^{\sigma+i\infty} F(s)e^{st} d\omega,$$

so

$$f(0) = \frac{1}{2\pi} \int_{\sigma-i\infty}^{\sigma+i\infty} F(s)e^{0} d\omega = \frac{1}{2\pi} \int_{\sigma-i\infty}^{\sigma+i\infty} F(s) d\omega, \qquad (3.9)$$

which is proportional to the area under the curve in a plot of $F(s)$ vs. $\omega$ at a fixed value of $\sigma$.

Note also that compressing a time-domain function along the time axis doesn't change its value at $t = 0$, so the area under the frequency-domain curve must remain the same. But the greater frequency extent of a frequency-domain function stretched out by factor $a$ would result in a larger area under the curve unless the amplitude is reduced by the same factor. This is why a factor of $1/a$ is needed outside $F(s/a)$ in Eq. 3.8.

Working Problem 3.3 at the end of this chapter will give you the opportunity to apply both the linearity and the scaling properties of the Laplace transform to the same time-domain function and compare the results.

## 3.4 Time-Domain Differentiation

One of the most useful characteristics of the Laplace transform in solving differential equations comes from the effect of taking the derivative of the time-domain function $f(t)$ prior to taking the Laplace transform. As you will see in Chapter 4, in many cases this property can be used to turn differential equations into algebraic equations, greatly simplifying the process of finding a solution.

The effect of differentiation prior to Laplace transformation can be seen by writing the unilateral Laplace-transform integral of the time derivative of $f(t)$ as

$$\mathcal{L}\left[\frac{df(t)}{dt}\right] = \int_{0-}^{\infty} \frac{df(t)}{dt} e^{-st} dt = \int_{0-}^{\infty} e^{-st} df = \lim_{\tau \to \infty} \int_{0-}^{\tau} e^{-st} df. \quad (3.10)$$

This integral can be simplified using integration by parts:

$$\int_{0}^{\tau} u\,dv = uv \bigg|_{0}^{\tau} - \int_{0}^{\tau} v\,du. \quad (3.11)$$

Applying this to Eq. 3.10 with

$$u = e^{-st} \qquad du = -se^{-st} dt \qquad v = f \qquad dv = df$$

gives

$$\mathcal{L}\left[\frac{df(t)}{dt}\right] = \lim_{\tau \to \infty} \left[ e^{-st} f(t) \bigg|_{0-}^{\tau} - \int_{0-}^{\tau} f(t)\left(-se^{-st}\right) dt \right]$$

$$= \lim_{\tau \to \infty} \left[ e^{-s(\tau)} f(\tau) - e^{-s(0)} f(0^{-}) + s \int_{0-}^{\infty} f(t) e^{-st} dt \right].$$

But $e^{-s(\tau)} f(\tau) = 0$ as $\tau \to \infty$, since the time-domain function $f(t)$ is of exponential order, and the last integral is just $F(s)$, the unilateral Laplace transform of $f(t)$. So

$$\mathcal{L}\left[\frac{df(t)}{dt}\right] = 0 - (1)f(0^{-}) + sF(s) = sF(s) - f(0^{-}). \quad (3.12)$$

This tells you that if $F(s)$ is the Laplace transform of $f(t)$, taking the derivative with respect to time of $f(t)$ prior to taking the Laplace transform has the effect of multiplying $F(s)$ by $s$, as long as you then subtract the value of $f(t)$ at time $t = 0^{-}$ (that is, the value of the function just before time $t = 0$) from the result.

To understand why that is true, remember that the result $F(s)$ of the Laplace transform tells you the amount of each basis function that adds in to the mix that synthesizes the modified time-domain function $f(t)u(t)e^{-\sigma t}$. So when you take the time derivative of $f(t)$, the Laplace transform of the result tells you the amount of each sinusoidal basis function in the mix that makes up the *change* in the modified time-domain function over time (specifically, the slope $df(t)/dt$ of the function $f(t)$ over time).

Remember also that the process of synthesizing a time-domain function from $F(s)$ involves multiplying $F(s)$ not only by the sinusoidal basis functions $e^{i\omega t}$, but also by the real exponential factor $e^{\sigma t}$, as described in Section 1.5. So when synthesizing the time-domain function $df/dt$ (representing the change of $f(t)$ over time), it's reasonable to expect that the change over time of the sinusoidal basis functions $e^{i\omega t}$ as well as the function $e^{\sigma t}$ will play a role.

That change over time is straightforward to assess mathematically: taking the time derivative brings down a factor of $i\omega$ from $e^{i\omega t}$ and a factor of $\sigma$ from $e^{\sigma t}$. You may find it instructive to consider why that happens, and to do that, recall that taking the time derivative of a function at any time is equivalent to finding the slope in the graph of the function at that time. The process of finding the slope for sinusoidal and exponential functions is illustrated in the "Derivatives" document on this book's website.

The key point is not simply that the process of taking the time derivative of the complex-exponential function $e^{i\omega t}$ brings a factor of $i\omega$ down from the exponent, but also that this factor is multiplied by the function itself. Hence the change in a complex-exponential basis function over time is just $i\omega$ times that same basis function.

Likewise, the factor of $\sigma$ brought down from the exponent by the time derivative of the real exponential function $e^{\sigma t}$ is also multiplied by the function itself.

In other words, in determining the change over time of the basis functions $e^{i\omega t}$ and the real exponential function $e^{\sigma t}$ that are used to synthesize the time-domain function $f(t)$, you find that the same basis functions are still present, scaled by $i\omega$, and the same real exponential is still present, scaled by $\sigma$.

With that understanding, it seems reasonable that the Laplace transform of the time derivative of $f(t)$ involves the same complex frequency-domain function $F(s)$, scaled by $\sigma$ to account for the change in $e^{\sigma t}$ over time, and by $i\omega$ to account for the change in the basis functions $e^{i\omega t}$ over time. And since $s = \sigma + i\omega$, multiplying $F(s)$ by $s$ accomplishes the required scaling.

This analysis hasn't addressed the $f(0^-)$ that is subtracted from $sF(s)$ in Eq. 3.12. If you look up the Fourier transform or the bilateral (two-sided) Laplace transform of the derivative of a time-domain function, you'll see that this term is not present in either of those transforms. The reason for that can be seen by taking another look at the derivation of Eq. 3.12 above. As you can see, the $f(0^-)$ term originates from the insertion of the lower limit in the $uv$ product when employing integration by parts. In the Fourier and bilateral Laplace transforms, the lower integration limit is $-\infty$, and the term involving

$f(-\infty)e^{-\infty}$ vanishes for integrable functions. But in the unilateral Laplace transform, the term $-f(0^-)e^0$ may be nonzero, which is why it appears in Eq. 3.12.

You should also note that versions of Eq. 3.12 for higher-order derivatives also exist. The version for second-order time derivatives looks like this:

$$\mathcal{L}\left[\frac{d^2 f(t)}{dt^2}\right] = s^2 F(s) - sf(0^-) - \left.\frac{df(t)}{dt}\right|_{t=0^-}, \qquad (3.13)$$

and for third-order derivatives

$$\mathcal{L}\left[\frac{d^3 f(t)}{dt^3}\right] = s^3 F(s) - s^2 f(0^-) - s\left.\frac{df(t)}{dt}\right|_{t=0^-} - \left.\frac{d^2 f(t)}{dt^2}\right|_{t=0^-}. \qquad (3.14)$$

For time derivatives of fourth order and higher, the derivative property can be written as

$$\mathcal{L}\left[\frac{d^n f(t)}{dt^n}\right] = s^n F(s) - s^{n-1} f(0^-) - s^{n-2}\left.\frac{df(t)}{dt}\right|_{t=0^-} - \cdots$$

$$- s\left.\frac{d^{n-2} f(t)}{dt^{n-2}}\right|_{t=0^-} - \left.\frac{d^{n-1} f(t)}{dt^{n-1}}\right|_{t=0^-}, \qquad (3.15)$$

in which $n$ represents the order of the derivative taken of $f(t)$ prior to performing the Laplace transformation.

If you are wondering about the dimensional compatibility of the terms in these equations, remember that the Laplace transform $\mathcal{L}[f(t)]$ has dimensions of $f(t)$ multiplied by time, as explained in Chapter 1. That means that the transform $\mathcal{L}\left[\frac{df(t)}{dt}\right]$ has dimensions of $(f(t)/\text{time}) \times \text{time}$, which are the same dimensions as $f(t)$. Likewise, the dimensions of the product $sF(s)$ are the same as those of $f(t)$, since $s$ has dimensions of $1/\text{time}$. Hence all of the terms in the equation for the derivative property of the Laplace transform have the dimensions of $f(t)$.

You can apply the time-domain derivative property of the Laplace transform to sinusoidal, exponential, and $t^n$ functions in Problem 3.4 at the end of this chapter.

## 3.5 Time-Domain Integration

If you worked through the development in the previous section, the approach to finding the Laplace transform of the integral of the time-domain function $f(t)$

over time should look familiar to you. Start by writing the integral of $f(t)$ over the time period from $0^-$ to $t$ as

$$\text{integral of } f(t) \text{ over time} = \int_{0^-}^{t} f(\tau)d\tau, \tag{3.16}$$

in which the integration variable is called $\tau$ in order to distinguish it from the time variable $t$ of the Laplace transform. Taking the Laplace transform of this integral looks like this:

$$\mathcal{L}\left[\int_{0^-}^{t} f(\tau)d\tau\right] = \int_{0^-}^{\infty}\left[\int_{0^-}^{t} f(\tau)d\tau\right]e^{-st}dt. \tag{3.17}$$

As in the case of the time derivative, this can be simplified using integration by parts, with variables

$$u = \int_{0^-}^{t} f(\tau)d\tau \qquad du = f(t)dt \qquad dv = e^{-st}dt \qquad v = -\frac{1}{s}e^{-st}.$$

Inserting these expressions into Eq. 3.17 and using integration by parts (Eq. 3.11) gives

$$\mathcal{L}\left[\int_{0^-}^{t} f(\tau)d\tau\right] = \left[\left(\int_{0^-}^{t} f(\tau)d\tau\right)\left(-\frac{1}{s}e^{-st}\right)\right]\Bigg|_{0^-}^{\infty} - \int_{0^-}^{\infty}\left(-\frac{1}{s}e^{-st}\right)f(t)dt$$

$$= \left[\left(\int_{0^-}^{\infty} f(\tau)d\tau\right)\left(-\frac{1}{s}e^{-s(\infty)}\right)\right]$$

$$- \left[\left(\int_{0^-}^{0^-} f(\tau)d\tau\right)\left(-\frac{1}{s}e^{-s(0)}\right)\right] + \frac{1}{s}\int_{0^-}^{\infty} e^{-st}f(t)dt.$$

The expression in the first set of square brackets in the final equation is zero as long as the integral of $f(t)$ is of exponential order, since the exponential term goes to zero faster than the integral increases. The expression in the second set of square brackets is also zero, since the integration upper and lower limits are the same. That leaves only the third term, in which the integral is $F(s)$, the unilateral Laplace transform of $f(t)$. Hence

$$\mathcal{L}\left[\int_{0^-}^{t} f(u)du\right] = \frac{1}{s}\int_{0^-}^{\infty} e^{-st}f(t)dt = \frac{F(s)}{s}. \tag{3.18}$$

So the Laplace transform of the integral of the time-domain function $f(t)$ over time $t$ is just $F(s)$, the Laplace transform of $f(t)$, divided by $s$. You may have expected this result, since the Laplace transform of the time derivative of $f(t)$ involves $F(s)$ times $s$, and integrating a function over time is the inverse process of taking the derivative of that function with respect to time.

That intuition is helpful, and the result shown in Eq. 3.18 can be understood using similar logic to that of the previous section. In this case, instead of considering the change of the time-domain function $f(t)$ over time, think about the meaning of the integral of Eq. 3.16. Whereas the time derivative is equivalent to the slope of the graph of $f(t)$ vs. time, the integral over time is the area under the curve of $f(t)$ in that graph. So when you take the Laplace transform of the integral of $f(t)$ over time, the result tells you the amount of each of the sinusoidal basis functions that mix together to provide the area under the curve of the modified function $f(t)u(t)e^{-\sigma t}$.

And since that mixing process, which is the synthesis done by the inverse Laplace transform, involves multiplication of $F(s)$ by the complex exponential basis functions $e^{i\omega t}$ and also by the real exponential function $e^{\sigma t}$, it is reasonable to expect that the area under the curve of the function $e^{(\sigma+i\omega)t}$ will play a role. That area is given by the integral of $e^{(\sigma+i\omega)t}$ over time interval $t$, and the result of that integration is just the function itself, $e^{(\sigma+i\omega)t}$, scaled by a factor of $1/(\sigma+i\omega)$. So it makes sense that the Laplace transform of the area under $f(t)$ is $F(s)$, the Laplace transform of $f(t)$, divided by $s = \sigma+i\omega$.

Problem 3.5 at the end of this chapter contains several functions to which you can practice applying the time-integration property of the Laplace transform.

## 3.6 Multiplication and Division of $f(t)$ by $t$

Although the Laplace transform of many functions can be determined using an integral equation such as Eq. 1.2 in Section 1.1, it is often quicker to use a table of known Laplace transforms for basic functions and then apply the rules for various combinations of those functions. One type of combination that occurs quite frequently in some applications of the Laplace transform takes the form of a function $f(t)$ that is multiplied or divided by the time parameter $t$, forming $tf(t)$ or $f(t)/t$. In this section, you will see that knowing $F(s)$, the Laplace transform of $f(t)$, allows you to find the Laplace transform of $tf(t)$ by taking the negative of the derivative of $F(s)$ with respect to $s$, and to find the Laplace transform of $f(t)/t$ by taking the integral of $F(s)$ over $s$.

### Multiplication of $f(t)$ by $t$

Showing that the Laplace transform of $tf(t)$ is related to the derivative of $F(s)$ with respect to $s$ is straightforward. Start by writing the integral equation for $F(s)$ (Eq. 1.2) and then taking the derivative with respect to $s$ of both sides:

$$F(s) = \mathcal{L}[f(t)] = \int_{0-}^{\infty} f(t)e^{-st}\,dt$$

$$\frac{d[F(s)]}{ds} = \frac{d}{ds}\left[\int_{0-}^{\infty} f(t)e^{-st}\,dt\right].$$

The derivative with respect to $s$ is a linear operation that can be moved inside the integral over time, so

$$\frac{d[F(s)]}{ds} = \int_{0-}^{\infty} f(t)\frac{d(e^{-st})}{ds}\,dt = \int_{0-}^{\infty} f(t)(-te^{-st})\,dt$$

$$= -\int_{0-}^{\infty} [tf(t)]e^{-st}\,dt = -\mathcal{L}[tf(t)].$$

Thus

$$\mathcal{L}[tf(t)] = -\frac{d[F(s)]}{ds}. \tag{3.19}$$

If you are concerned about taking the derivative of $F(s)$ with respect to a complex variable ($s$), you needn't be. Once you have an expression for $F(s)$ in which $s$ appears, you can treat $s$ like any other variable for the purposes of taking the derivative of $F(s)$.

That is the mathematical reasoning behind this property of the Laplace transform, and Eq. 3.19 can be extremely useful in finding the Laplace transform of a function of the form $tf(t)$. To get an intuitive sense of why this relationship is true, remember that the generalized frequency-domain function $F(s)$ is the result of forming the inner product between the modified function $f(t)u(t)e^{-\sigma t}$ and testing functions, which in this case are the sinusoidal basis functions $e^{i\omega t}$. That inner product involves multiplying the modified function by the complex conjugate of the basis functions (that is, $e^{-i\omega t}$) and integrating the result over time. The result of that process, $F(s)$, tells you the amount of each basis function present in the modified function $f(t)u(t)e^{-\sigma t}$ for a given value of $s = \sigma + i\omega$.

But the modifying function $e^{-\sigma t}$ and the complex conjugate of the basis function $e^{-i\omega t}$ both change as $s$ changes, because a change in the real part of $s$ changes $\sigma$, and a change in the imaginary part of $s$ changes $i\omega$. And here's the key point: if you ask how quickly those functions change as $s$ changes, the answer is that the rate of change of both functions is $-t$ times the functions themselves. That is, the rate of change of $e^{-\sigma t}$ with $\sigma$ is $-t$ times $e^{-\sigma t}$, and the rate of change of $e^{-i\omega t}$ with $i\omega$ is $-t$ times $e^{-i\omega t}$. If you use that rate of change (with $s$) of the two functions under the time integral on the right side of the Laplace transform equation (Eq. 1.2), you get the rate of change (with $s$) of the function $F(s)$ on the left side of the equation.

Of course, seeing the mathematical proof and having an intuitive feel for the result are both helpful, but if you still feel a bit uncertain about this relationship, you may find it helpful to see the time-domain function $tf(t)$ synthesized from the $s$-domain function $-dF(s)/ds$. You can see an example of that synthesis on this book's website.

There is also a rule for cases in which the time-domain function is a higher power of $t$ times $f(t)$, which you can find by repeated application of the development shown for $tf(t)$. As you might expect, that rule relates higher powers of $t$ to higher-order derivatives of $F(s)$:

$$\mathcal{L}[t^n f(t)] = (-1)^n \frac{d^n}{ds^n} F(s), \tag{3.20}$$

in which the factor of $-1$ is taken to the power of $n$ (the power to which $t$ is raised) because each application of the derivative with respect to $s$ brings down an additional factor of $-t$.

## Division of $f(t)$ by $t$

Deriving the property that relates the integral of $F(s)$ to the time-domain function $f(t)/t$ takes a few more steps than in the case of multiplication by $t$, but none of those steps is difficult. In this case, start by writing the Laplace transform equation for the function $g(t) = f(t)/t$:

$$\mathcal{L}[g(t)] = \int_{0^-}^{\infty} \frac{f(t)}{t} e^{-st} dt = G(s).$$

Now use the result from the previous section that the Laplace transform of $tf(t)$ is $-dF(s)/ds$, applied to $g(t)$:

$$\mathcal{L}[tg(t)] = \int_{0^-}^{\infty} [tg(t)] e^{-st} dt = -\frac{dG(s)}{ds}.$$

But $tg(t) = f(t)$, so

$$-\frac{dG(s)}{ds} = \int_{0^-}^{\infty} [tg(t)] e^{-st} dt = \int_{0^-}^{\infty} [f(t)] e^{-st} dt = F(s).$$

This equation relates the change in $G(s)$, the Laplace transform of $g(t) = f(t)/t$, to $F(s)$, and the Laplace transform of $f(t)$. To find an expression for $G(s)$ itself, it is necessary to integrate both sides of this equation, and setting the lower limit of that integration to $s$ (for reasons that will become clear below) and the upper limit to $\infty$ gives

$$\int_s^\infty \left[ -\frac{dG(u)}{du} \right] du = \int_s^\infty F(u) du$$

or

$$-[G(\infty) - G[s)] = \int_s^\infty F(u) du,$$

in which the generalized-frequency variable $u$ is introduced to differentiate the lower limit of integration from the integration variable.

Since all functions that are Laplace transforms such as $G(s)$ must go to zero as $s$ goes to $\infty$, $G(\infty) = 0$, which means

$$G(s) = \mathcal{L}[g(t)] = \int_s^\infty F(u) du.$$

And since $g(t) = f(t)/t$, this is

$$\mathcal{L}\left[ \frac{f(t)}{t} \right] = \int_s^\infty F(u) du. \tag{3.21}$$

Using logic similar to that in the case of multiplication of $f(t)$ by $t$, an intuitive sense of why this is true can be gained by once again recalling that $F(s)$ is given by the inner product between the modified time-domain function and the sinusoidal basis functions of the Laplace transform. But in this case, instead of asking how the real and complex exponential functions (and therefore $F(s)$) change with $s$, consider the area under the curve of those functions when plotted against $s$. That area is found by integrating the functions over $s$, and for the function $e^{-st}$, the integral between the limits of $s$ and $\infty$ results in a factor of $1/t$ times the function $e^{-st}$ itself when evaluated at the limits (this is why the lower limit of integration was set to $s$). So finding the area under the curve of the $e^{-st}$ function on the right side of the Laplace transform equation (Eq. 1.2) results in the transform of $f(t)/t$, and doing the same on the left side results in the integral of the frequency-domain function $F(u)$ between $s$ and $\infty$.

At the end of this chapter, Problem 3.6 provides several functions to which you can apply the Laplace-transform properties of multiplication and division by $t$.

## 3.7 Transform of Periodic Functions

This property of the Laplace transform applies to periodic time-domain functions – that is, functions for which the values of the function repeat themselves at later times, such as sinusoidal functions or a train of identical pulses. If $f(t)$ is a periodic function, then

$$f(t) = f(t + T), \qquad (3.22)$$

in which $T$ represents the period of the function.

For any periodic function that is continuous at times $t = 0$, $t = T$, and all other multiples of $T$ (that is, a function that does not have discontinuities at the beginning or end of each period), the unilateral Laplace transform can be found by performing the usual integration of $f(t)e^{-st}$, but rather than integrating from $t = 0$ to $t = \infty$, the upper limit can be set equal to $T$. In other words, the integral can be taken over a single period, as long as you then divide the results of that single-period integration by a factor of $1 - e^{-sT}$. That makes the equation for the Laplace transform of a periodic function look like this:

$$\mathcal{L}[f(t)] = \frac{\int_0^T f(t)e^{-st}dt}{1 - e^{-sT}}. \qquad (3.23)$$

To see why this works, break the integral in the Laplace transform (Eq. 1.2) into two parts, with the first part from $t = 0$ to $t = T$, and the second part from $t = T$ to $t = \infty$:

$$\mathcal{L}[f(t)] = \int_0^\infty f(t)e^{-st}dt = \int_{t=0}^{t=T} f(t)e^{-st}dt + \int_{t=T}^{t=\infty} f(t)e^{-st}dt.$$

Now in the second integral let $t = \tau + T$ and $dt = d\tau$:

$$\mathcal{L}[f(t)] = \int_{t=0}^{t=T} f(t)e^{-st}dt + \int_{\tau+T=T}^{\tau+T=\infty} f(\tau + T)e^{-s(\tau+T)}d\tau$$

$$= \int_{t=0}^{t=T} f(t)e^{-st}dt + \int_{\tau=0}^{\tau=\infty} f(\tau + T)e^{-s(\tau+T)}d\tau$$

$$= \int_{t=0}^{t=T} f(t)e^{-st}dt + e^{-sT}\int_{\tau=0}^{\tau=\infty} f(\tau + T)e^{-s\tau}d\tau.$$

Since $f(t)$ is a periodic function with period $T$, Eq. 3.22 tells you that $f(\tau+T)$ must equal $f(\tau)$, so

$$\mathcal{L}[f(t)] = \int_{t=0}^{t=T} f(t)e^{-st}dt + e^{-sT}\int_{\tau=0}^{\tau=\infty} f(\tau)e^{-s\tau}d\tau.$$

The second integral in this equation is the Laplace transform of $f(t)$ (the name of the integration variable doesn't affect the relationship), so

$$\mathcal{L}[f(t)] = \int_0^T f(t)e^{-st}dt + e^{-sT}\mathcal{L}[f(t)]$$

or

$$\mathcal{L}[f(t)] - e^{-sT}\mathcal{L}[f(t)] = \int_0^T f(t)e^{-st}dt$$

$$\mathcal{L}[f(t)](1 - e^{-sT}) = \int_0^T f(t)e^{-st}dt$$

$$\mathcal{L}[f(t)] = \frac{\int_0^T f(t)e^{-st}dt}{1 - e^{-sT}}$$

in agreement with Eq. 3.23. This result follows mathematically, but why is it necessary to integrate over only one period, and what does that term in the denominator mean?

To understand the answer to those questions, it may help you to think of the integral of all the repetitions of the function $f(t)$ after the first period as the sum of all of the integrals taken over each period $T$:

$$\int_T^\infty f(t)e^{-st}dt = \int_T^{2T} f(t)e^{-st}dt + \int_{2T}^{3T} f(t)e^{-st}dt$$

$$+ \int_{3T}^{4T} f(t)e^{-st}dt + \ldots$$

and so on for an infinite number of periods.

The sum of those contributions doesn't grow to infinity because $e^{-st}$ includes a factor of $e^{-\sigma t}$, so, as long as $\sigma$ is sufficiently large and $f(t)$ is piecewise continuous and of exponential order, the integral converges even with an infinite number of periods. But what do all those other terms add up to?

To determine that, start by looking at the term for the second period ($t = T$ to $t = 2T$). If you make the substitution $t = \tau + T$ and $dt = d\tau$, that term becomes

$$\int_{t=T}^{t=2T} f(t)e^{-st}dt = \int_{\tau+T=T}^{\tau+T=2T} f(\tau + T)e^{-s(\tau+T)}d\tau$$

$$= \int_{\tau=0}^{\tau=T} f(\tau + T)e^{-s\tau}e^{-sT}d\tau$$

$$= e^{-sT}\int_{\tau=0}^{\tau=T} f(\tau + T)e^{-s\tau}d\tau.$$

But $f(\tau + T) = f(\tau)$, so this is

$$\int_{t=T}^{t=2T} f(t)e^{-st}dt = e^{-sT}\int_{\tau=0}^{\tau=T} f(\tau)e^{-s\tau}d\tau = e^{-sT}\int_{t=0}^{t=T} f(t)e^{-st}dt.$$

Adding this to the integral over the first period gives

$$\int_0^T f(t)e^{-st}dt + e^{-sT}\int_0^T f(t)e^{-st}dt = (1+e^{-sT})\int_0^T f(t)e^{-st}dt.$$

The same process for the integral over the third period ($t = 2T$ to $t = 3T$), with the substitution $t = \tau + 2T$ and $dt = d\tau$, gives

$$\int_{t=2T}^{t=3T} f(t)e^{-st}dt = e^{-2sT}\int_{t=0}^{t=T} f(t)e^{-st}dt$$

and adding this to the integrals over the first two periods gives

$$\int_0^T f(t)e^{-st}dt + e^{-sT}\int_0^T f(t)e^{-st}dt + e^{-2sT}\int_0^T f(t)e^{-st}dt$$

$$= (1+e^{-sT}+e^{-2sT})\int_0^T f(t)e^{-st}dt.$$

Continuing this same process to later periods gives

$$\int_0^\infty f(t)e^{-st}dt = (1+e^{-sT}+e^{-2sT}+e^{-3sT}+\ldots)\int_0^T f(t)e^{-st}dt.$$

So the integral over each period gives the same contribution as the integral over the first period multiplied by $(e^{-sT})^{(n-1)}$, where $n$ is the number of the period (two for the second period, three for the third period, and so on).

The final step is to realize that for $|x| < 1$

$$1+x+x^2+x^3+\cdots = \frac{1}{1-x}$$

and applying this with $x = e^{-sT}$ gives

$$\int_0^\infty f(t)e^{-st}dt = \frac{1}{1-e^{-sT}}\int_0^T f(t)e^{-st}dt$$

in accordance with Eq. 3.23.

So it's the fact that each cycle of the periodic function $f(t)$ contributes the same amount (scaled by $(e^{sT})^{(n-1)}$) to the Laplace transform integral that leads to this property.

You can apply this property of the Laplace transform to the periodic function provided in Problem 3.7 at the end of this chapter, and you can check your result in the interactive solutions on this book's website.

## 3.8 Convolution

The mathematical operation of convolution is a form of multiplication of two functions that has important applications in signal processing, statistics, probability, image processing, and other fields in mathematics, physics, and engineering. For two causal time-domain functions $f(t)$ and $g(t)$ (that is, two functions which both have zero amplitude for all negative time), the convolution $(f * g)(t)$ is defined as

$$(f * g)(t) = \int_{\tau=0}^{\tau=t} f(\tau)g(t - \tau)d\tau. \tag{3.24}$$

Note that the convolution result $f * g$ is a function of time $t$, and that $\tau$ is a time-domain variable of integration which plays no role in the result, since it disappears when the integration limits are inserted.

If you haven't encountered convolution in your studies, or if you're uncertain about the process, you may find it helpful to take a look at the short tutorial called "Convolution" on this book's website.

And if you have seen convolution before, especially in the context of Fourier transforms, you may be wondering about the limits of integration in Eq. 3.24, since in some applications it is common to define the convolution integral over all time, from $t = -\infty$ to $t = \infty$. The reason that a lower limit of $\tau = 0$ is used in the definition above is that a causal function such as $f(t)$ must have an amplitude of zero for all negative time, so the smallest value of $\tau$ for which $f(\tau)$ can contribute to the convolution is $f(0)$. Likewise, the upper limit of $\tau = t$ is used because if $\tau$ were allowed to become greater than $t$, the argument of $g(t-\tau)$ would be negative, and since $g(t)$ is also causal, this function cannot contribute to the convolution if its argument is less than zero.

So exactly what happens between those limits of $\tau$? Here are the major steps represented by Eq. 3.24: one of the two functions, $g(\tau)$ in this case, is reversed in time, making $g(-\tau)$. This function is then offset in time by amount $t$, making $g(t - \tau)$, multiplied by the other function (which is $f(\tau)$ in this case), and the multiplication products are added up (that's the integral over $\tau$). This reverse, shift, multiply, and accumulate process gives one point (at one time $t$) in the function that is the convolution $(f * g)(t)$. Then as time passes and $t$ increases, the function $g(t - \tau)$ "slides past" the function $f(\tau)$. As that happens, the point-by-point multiplication of $g(t - \tau)$ and $f(\tau)$ is repeated, and the results are accumulated for each value of $t$. So at each position of the shifted function $g(t - \tau)$, it is the overlap of the two functions that contributes to the convolution.

What does the convolution of two functions tell you? There are several ways to interpret the convolution result, but the common thread is that the convolution process is a form of weighted averaging of one of the functions over time, with the values of the other function providing the weights that are applied in the averaging process. That is why you may have seen convolution described as a blending process in which one of the functions is used to "smear out" the other function. That doesn't sound particularly useful, but in fact the convolution process can accomplish much more than blurring.

One reason that convolution is immensely useful is that the Laplace transform of the convolution between two time-domain functions gives the same result as simply multiplying the Laplace transforms of those two functions point by point. For example, if the Laplace transform of $f(t)$ is $F(s)$ and the Laplace transform of $g(t)$ is $G(s)$

$$F(s) = \int_0^\infty f(t)e^{-st}\,dt \qquad\qquad G(s) = \int_0^\infty g(t)e^{-st}\,dt,$$

then the Laplace transform of the convolution of $f(t)$ and $g(t)$ is the product of $F(s)$ and $G(s)$:

$$\mathcal{L}[(f * g)(t)] = \mathcal{L}[f(t)]\mathcal{L}[g(t)] = F(s)G(s). \tag{3.25}$$

To see why that's true, start by writing the product $F(s)G(s)$ as

$$F(s)G(s) = \int_{\tau=0}^{\tau=\infty} f(\tau)e^{-s\tau}\,d\tau \int_{\alpha=0}^{\alpha=\infty} g(\alpha)e^{-s\alpha}\,d\alpha, \tag{3.26}$$

in which separate variables $\tau$ (representing the time variable in $f$) and $\alpha$ (representing the time variable in $g$) are used in order to distinguish the integration variable for each function. The relevant time variables $\tau$ and $\alpha$ are explicitly shown in the integration limits to help you keep track of the limits as these integrals change position in the equations and as a change of variables is made.

Since the integral of $g(\alpha)$ has no dependence on $\tau$, that integral can be moved inside the $\tau$ integral, and the terms $f(\tau)$ and $e^{-s\tau}$ can be moved inside the $\alpha$ integral for the same reason:

$$F(s)G(s) = \int_{\tau=0}^{\tau=\infty} f(\tau)e^{-s\tau} \left[ \int_{\alpha=0}^{\alpha=\infty} g(\alpha)e^{-s\alpha}\,d\alpha \right] d\tau$$

$$= \int_{\tau=0}^{\tau=\infty} \int_{\alpha=0}^{\alpha=\infty} f(\tau)g(\alpha)e^{-s(\tau+\alpha)}\,d\alpha\,d\tau.$$

Now define the following variables:

$$t = \tau + \alpha \qquad\qquad \alpha = t - \tau \qquad\qquad d\alpha = dt$$

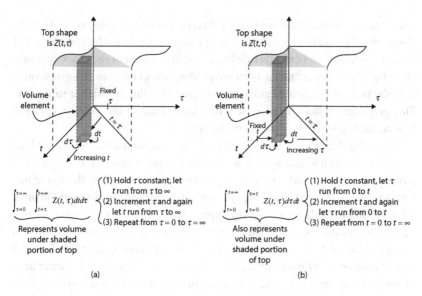

Figure 3.2 Changing the order of integration in a double integral of the function $Z(t, \tau)$.

and insert these expressions for $\alpha$ and $d\alpha$ into the equation for $F(s)G(s)$. That gives

$$F(s)G(s) = \int_{\tau=0}^{\tau=\infty} \int_{t-\tau=0}^{t-\tau=\infty} f(\tau)g(t-\tau)e^{-s(\tau+t-\tau)} dt d\tau$$

$$= \int_{\tau=0}^{\tau=\infty} \int_{t=\tau}^{t=\infty} f(\tau)g(t-\tau)e^{-st} dt d\tau. \qquad (3.27)$$

To relate this equation to the Laplace transform of the convolution of $f(t)$ and $g(t)$, it is necessary to have the integration over $t$ on the outside and the integration over $\tau$ on the inside – in other words, to switch the order of these two integrals.

To see how that's done, recall that the double integration of a function of two variables, such as $Z(t, \tau) = f(\tau)g(t-\tau)e^{-st}$, can be thought of as the process of finding the volume between the function $Z(t, \tau)$ and the $t, \tau$ plane over some range of values of $t$ and $\tau$, as shown in Fig. 3.2. In both Fig. 3.2a and Fig. 3.2b, that total volume is found using a volume element of area $dt d\tau$ and height $Z(t, \tau)$, and the order of integration in the double integral determines how that volume element is swept over the region of interest.

Figure 3.2a illustrates how this works with the $\tau$ integral on the outside and the $t$ integral on the inside. In that case, the total volume is determined by sweeping the volume element in the $t$ direction at a fixed value of $\tau$. The line along which the volume element is swept begins at $t = \tau$ (the lower limit of the inner integral) and extends to infinity (the upper limit of the inner integral). Once the volume element has been swept along the entire line at a given value of $\tau$, the process is repeated along another (parallel) line at a higher value of $\tau$. This procedure is done from $\tau = 0$ (the lower limit of the outer integral) to $\tau = \infty$ (the upper limit of the outer integral).

Switching the order of integration puts the $t$ integral on the outside and the $\tau$ integral on the inside, and in that case the volume element is swept in the $\tau$ direction while $t$ is held constant, as shown in Fig. 3.2b. Now the line along which the volume element is swept begins at $\tau = 0$ (the lower limit of the inner integral) and extends to $\tau = t$ (the upper limit of the inner integral). Once again, after sweeping the volume element along the entire line (this time at a given value of $t$), the process is repeated along another (parallel) line at a higher value of $t$. In this case the procedure is done from $t = 0$ (the lower limit of the outer integral) to $t = \infty$ (the upper limit of the outer integral).

The key point is that the result is the same for both methods of determining the volume between the $Z(t, \tau)$ function and the $t, \tau$ plane (that is, the value of the double integral), although the limits of integration are different in the two approaches.

After switching the order of integration in Eq. 3.27, the product of $F(s)$ and $G(s)$ looks like this:

$$F(s)G(s) = \int_{\tau=0}^{\tau=\infty} \int_{t=\tau}^{t=\infty} f(\tau)g(t-\tau)e^{-st}dt\,d\tau$$

$$= \int_{t=0}^{t=\infty} \int_{\tau=0}^{\tau=t} f(\tau)g(t-\tau)e^{-st}d\tau\,dt$$

$$= \int_{t=0}^{t=\infty} \left[ \int_{\tau=0}^{\tau=t} f(\tau)g(t-\tau)d\tau \right] e^{-st}dt.$$

Comparing the inner integral to the convolution equation shown in Eq. 3.24, you can see that

$$F(s)G(s) = \int_{t=0}^{t=\infty} \left[ \int_{\tau=0}^{\tau=t} f(\tau)g(t-\tau)d\tau \right] e^{-st}dt$$

$$= \int_{t=0}^{t=\infty} \left[ (f * g)(t) \right] e^{-st}dt,$$

which is just the Laplace transform of $(f * g)(t)$. Thus

$$F(s)G(s) = \int_{t=0}^{t=\infty} \left[ (f * g)(t) \right] e^{-st} dt = \mathcal{L}[(f * g)(t)], \qquad (3.28)$$

in agreement with Eq. 3.25 above.

So that is the mathematical proof that the Laplace transform of the convolution between two causal time-domain functions is the same as the product of the Laplace transforms of those functions. To gain an intuitive understanding of why this is true, it may help to remember that the time-domain functions $f(t)$ and $g(t)$ can be represented as the weighted superposition of basis functions, with the weights of those basis functions provided by the values of $F(s)$ and $G(s)$. From that perspective, the shift-and-multiply process of convolution is a form of polynomial multiplication, in which each term of one function is multiplied by all the terms of the other function. But in the Laplace transform, the basis functions that make up the modified time-domain functions $f(t)u(t)e^{-\sigma t}$ and $g(t)u(t)e^{-\sigma t}$ are orthogonal sinusoids, which means that the only terms that survive the process of multiplication and integration over time are those terms that have identical frequencies.

Hence it makes sense that the Laplace transform of those terms gives the point-by-point multiplication $F(s)G(s)$, in which every value of $F(s)$ is multiplied by the value of $G(s)$ for the same complex frequency $s$.

You can see how the convolution property of the Laplace transform works for two exponential functions in Problem 3.8 at the end of this chapter.

## 3.9 Initial- and Final-Value Theorems

The initial- and final-value theorems of the Laplace transform relate the behavior of the time-domain function $f(t)$ to the behavior of the $s$-domain function $F(s)$ at limiting values of $t$ and $s$. As its name implies, the initial-value theorem relates $f(t)$ to $F(s)$ for small values of time (and large values of $s$), while the final-value theorem pertains to large values of time (and small values of $s$). Both of these theorems can be derived from the Laplace-transform derivative characteristic discussed in Section 3.4, which says

$$\mathcal{L}\left[ \frac{df(t)}{dt} \right] = sF(s) - f(0^-), \qquad (3.12)$$

so if you're not clear on the meaning of this equation, this would be a good time to review that section.

## Initial-Value Theorem

The initial-value theorem tells you that the limit of the time-domain function $f(t)$ at very small (positive) values of $t$ equals the limit of $sF(s)$ at very large values of $s$:

$$\lim_{t\to0^+}[f(t)] = \lim_{s\to\infty}[sF(s)] \tag{3.29}$$

as long as $f(t)$ is of exponential order and the limits exist. In this equation, the limit as $s \to \infty$ is taken along the positive real axis.

To see the mathematics leading to the initial-value theorem, start by taking the limit of both sides of Eq. 3.12 as $s \to \infty$. To properly account for the possibility of a discontinuity in $f(t)$ at $t = 0$, for this theorem it is important to explicitly show that the unilateral Laplace transform integral lower limit is $0^-$:

$$\lim_{s\to\infty}\mathcal{L}\left[\frac{df(t)}{dt}\right] = \lim_{s\to\infty}[sF(s) - f(0^-)],$$

$$\lim_{s\to\infty}\int_{0^-}^{\infty}\frac{df(t)}{dt}e^{-st}dt = \lim_{s\to\infty}[sF(s)] - f(0^-),$$

$$\lim_{s\to\infty}\int_{0^-}^{\infty}e^{-st}df = \lim_{s\to\infty}[sF(s)] - f(0^-).$$

Now divide the integral into two portions, the first from just before $t = 0$ (that is, $0^-$) to just after $t = 0$ ($0^+$), and the second from $t = 0^+$ to $t = \infty$. That gives

$$\lim_{s\to\infty}\left[\int_{t=0^-}^{t=0^+}e^{-s(0)}df + \int_{t=0^+}^{t=\infty}e^{-st}df\right] = \lim_{s\to\infty}[sF(s)] - f(0^-).$$

The limit as $s$ approaches infinity of the second integral is zero, since $f(t)$ is piecewise continuous and of exponential order, so this is

$$\lim_{s\to\infty}\left[f(0^+) - f(0^-) + 0\right] = \lim_{s\to\infty}[sF(s)] - f(0^-),$$

$$f(0^+) - f(0^-) = \lim_{s\to\infty}[sF(s)] - f(0^-),$$

$$f(0^+) = \lim_{s\to\infty}[sF(s)].$$

Writing $f(0^+)$ as the limit of $f(t)$ as $t \to 0^+$ makes this

$$\lim_{t \to 0^+} f(t) = \lim_{s \to \infty} [sF(s)],$$

consistent with Eq. 3.29.

You may be wondering why this is true, that is, why the limit of $sF(s)$ as $s$ goes to $\infty$ has the same value as the time-domain function $f(t)$ at time $t = 0^+$. To understand this result, recall that large values of the real part of $s$ mean that the real exponential-decay factor $e^{-\sigma t}$ in the Laplace transform reduces the amplitude of the modified function $f(t)u(t)e^{-\sigma t}$ very quickly with time. That tends to suppress the behavior of $f(t)$ after time $t = 0^+$, which is why time-domain functions that include $f(t) = c$, $f(t) = e^{at}$, $f(t) = \cos(\omega_1 t)$, and $f(t) = \cosh(at)$ all have Laplace transforms $F(s)$ that vary as $1/s$ for large values of $s$. Hence the product $sF(s)$ approaches a constant value for each of those functions at large $s$, and the value of that constant is determined by the value of the function $f(t)$ at time $t = 0^+$.

Alternatively, time-domain functions such as $f(t) = \sin(\omega_1 t)$ and $f(t) = \sinh(at)$ have Laplace transforms that vary as $1/s^2$ for large values of $s$, so the product $sF(s)$ varies as $1/s$. That means that $sF(s)$ approaches zero as $s$ approaches $\infty$, which matches the value of these functions at time $t = 0^+$. Thus the initial-value theorem holds in these cases, as well.

## Final-Value Theorem

The final-value theorem relates the limit of the time-domain function $f(t)$ at large values of $t$ to the limit of $sF(s)$ at small values of $s$:

$$\lim_{t \to \infty} [f(t)] = \lim_{s \to 0} [sF(s)] \tag{3.30}$$

provided that the Laplace transform of $f(t)$ exists and that the limit of $f(t)$ as $t$ approaches infinity also exists. That means that the final-value theorem cannot be applied to pure sinusoidal functions, which oscillate forever and do not have a defined limit as $t$ approaches infinity, nor can it be applied to increasing exponentials, which grow without limit. But for any function $f(t)$ that has a Laplace transform and that reaches a limit as time increases, you can use the final-value theorem to determine the long-term (sometimes called the "steady-state") value of $f(t)$ by finding the value of $F(s)$ as $s$ goes to zero.

Like the initial-value theorem, the final-value theorem can be derived from the derivative property of the Laplace transform. In this case, take the limit of Eq. 3.12 as $s \to 0$:

$$\lim_{s \to 0} \mathcal{L}\left[\frac{df(t)}{dt}\right] = \lim_{s \to 0}[sF(s) - f(0^-)]$$

or

$$\lim_{s \to 0} \int_{0^-}^{\infty} \left[\frac{df(t)}{dt}\right] e^{-st} dt = \lim_{s \to 0}[sF(s)] - f(0^-).$$

Since $e^{st} \to 1$ as $s \to 0$, this can be written as

$$\int_{0^-}^{\infty} df(t) = \lim_{\tau \to \infty} \int_{0^-}^{\tau} df(t) = \lim_{s \to 0}[sF(s)] - f(0^-)$$

$$\lim_{\tau \to \infty} f(\tau) - f(0^-) = \lim_{s \to 0}[sF(s)] - f(0^-)$$

$$\lim_{\tau \to \infty} f(\tau) = \lim_{s \to 0}[sF(s)]$$

in agreement with Eq. 3.30.

You can see how the final-value theorem works (or doesn't apply) by considering the time-domain function $f(t) = Ae^{at}$ with values of the constant $a$ that are greater than zero, equal to zero, or less than zero. In the first case, with $a > 0$, $f(t)$ grows without limit as $t$ approaches infinity, which means that the final-value theorem cannot be applied.

But if $a = 0$, then $f(t)$ is constant in time, and the Laplace transform of $f(t)$ is $F(s) = A/s$ for all positive values of $s$. So in this case the limit of $sF(s) = s(A/s) = A$ as $s$ goes to 0, which equals the value of $f(t)$ at all values of $t$, so $A$ is the limit of $f(t)$ as $t$ approaches infinity.

The final-value theorem also applies to the function $f(t) = Ae^{at}$ when the value of $a$ is negative. In that case the limiting value of $f(t)$ as $t \to \infty$ is zero, and the Laplace transform gives $F(s) = A/(s - a)$. Hence $sF(s) = sA/(s - a)$, which also approaches zero as $s \to 0$.

Reading about the Laplace transforms of basic functions and the useful properties of the Laplace transform is a good first step, but if you want to develop a deep understanding of how to use the properties discussed in this chapter, you should make an effort to work through each of the problems in the next section. If you need some help as you do that, you'll find full interactive solutions for every problem on this book's website.

## 3.10 Problems

1. Use the linearity property of the unilateral Laplace transform and the examples of $F(s)$ for basic functions in Chapter 2 to find $F(s)$ for
$$f(t) = 5 - 2e^{\frac{t}{2}} + \tfrac{1}{3}\sin(6t) - 3t^4 + 8\sinh(0.2t).$$

2. Use the linearity, time-shift, and frequency-shift properties of the unilateral Laplace transform to find $F(s)$ for

   (a) $f(t) = 2e^{\frac{t-3}{2}} + \tfrac{1}{3}\sin(6t - 9) - 3(t - 2)^4 + 8\sinh(0.2t - 0.6)$

   (b) $f(t) = -5e^{-\frac{t}{3}} + e^t\cos\left(\frac{4t}{3}\right) + e^{\frac{t}{2}}\left(\frac{t}{2}\right)^2 - \tfrac{1}{3}\frac{\cosh(4t)}{e^{3t}}$.

3. Use the linearity property of the unilateral Laplace transform to find $F(s)$ for $f(t) = (2t)^3 + \left(\frac{t}{2}\right)^2$, then show that using the scaling property of Section 3.3 gives the same result.

4. Take the derivative of each of the following functions with respect to time, then find the unilateral Laplace transform of the resulting function and compare it to the result of using the time-derivative property of the Laplace transform on the original function:

   (a) $f(t) = \cos(\omega_1 t)$

   (b) $f(t) = e^{-2t}$

   (c) $f(t) = t^3$.

5. Integrate each of the following functions over time from 0 to $t$, then find the unilateral Laplace transform of the resulting function and compare it to the result of using the time-integration property of the Laplace transform on the original function:

   (a) $f(t) = \sin(\omega_1 t)$

   (b) $f(t) = 2e^{6t}$

   (c) $f(t) = 3t^2$.

6. Use the linearity property and the multiplication and division by $t$ properties of the unilateral Laplace transform to find $F(s)$ for

   (a) $f(t) = t^2\sin(4t) - t\cos(3t)$

   (b) $f(t) = \frac{\sin(\omega_1 t)}{t} + \frac{1-e^{-t}}{t}$

   (c) $f(t) = t\cosh(-t) - \frac{\sinh(2t)}{t}$.

7. Find the unilateral Laplace transform for the periodic triangular function $f(t) = t$ for $0 < t < 1$ sec and $f(t) = 2 - t$ for $1 < t < 2$ sec, if the period of this function is 4 seconds.

8. Find the convolution of the causal functions $f(t) = 3e^t$ and $g(t) = 2e^{-t}$ and show that

$$\mathcal{L}[f * g(t)] = F(s)G(s),$$

in which $F(s) = \mathcal{L}[f(t)]$ and $G(s) = \mathcal{L}[g(t)]$, in accordance with Eq. 3.25.

9. Find the unilateral Laplace transform for the time-domain function $f(t) = 5 - 2t + 3\sin(4t)e^{-2t}$ and show that the initial-value theorem (Eq. 3.29) holds in this case.

10. Find the unilateral Laplace transform for the time-domain function $f(t) = 4t^2e^{-3t} - 3 + e^{-t}\cosh(2t)$ and show that the final-value theorem (Eq. 3.30) holds in this case.

# 4

# Applications of the Laplace Transform

The previous three chapters were designed to help you understand the meaning and the method of the Laplace transform and its relation to the Fourier transform (Chapter 1), to show the Laplace transform of a few basic functions (Chapter 2), and to demonstrate some of the properties that make the Laplace transform useful (Chapter 3). In this chapter, you will see how to use the Laplace transform to solve problems in five different topics in physics and engineering. Those problems involve differential equations, so the first section of this chapter (Section 4.1) provides an introduction to the application of the Laplace transform to ordinary and partial differential equations. Once you have an understanding of the general concept of solving a differential equation by applying an integral transform, you can work through specific applications including mechanical oscillations (Section 4.2), electrical circuits (Section 4.3), heat flow (Section 4.4), waves (Section 4.5), and transmission lines (Section 4.6). Each of these applications has been chosen to illustrate a different aspect of using the Laplace transform to solve differential equations, so you may find them useful even if you have little interest in the specific subject matter. And as in every chapter, the final section (Section 4.7) of this chapter has a set of problems you can use to check your understanding of the concepts and mathematical techniques presented in this chapter.

## 4.1 Differential Equations

It's a common saying in physics and engineering that many of the important laws of nature are expressed as differential equations. That is true because natural laws often describe how quantities change over space and time, and the mathematical expressions for those spatial and temporal changes are ordinary derivatives for quantities depending on a single variable or partial derivatives for quantities depending on multiple variables. Those quantities can represent the condition of an object, the properties of a material, or the state of a process, and differential equations tell you how the spatial and temporal changes of those quantities relate to other quantities (or other changes).

Of course, if you want to know the value of the function that represents the quantity of interest, knowing how that function changes over a given distance or period of time is useful, but you also need to know its value at a specific location or time (and in some cases more than one location or time). Those known values are generally called "boundary conditions"; an "initial condition" is a boundary condition that occurs at a time or at a location that you have defined as zero.

So what is it about the Laplace transform that makes it so helpful in solving differential equations? Many of the properties discussed in Chapter 3 such as linearity, shifting, and scaling are important, but of special importance in this case is the property concerning derivatives. As you have seen if you've worked through Section 3.4, that property says that the Laplace transform of the derivative a time-domain function $f(t)$ with respect to time is related to the product of the complex frequency $s$ and the $s$-domain function $F(s)$ as well as the initial value of $f(t)$:

$$\mathcal{L}\left[\frac{df(t)}{dt}\right] = sF(s) - f(0^-), \tag{3.12}$$

and the version of this property for second-order time derivatives is

$$\mathcal{L}\left[\frac{d^2 f(t)}{dt^2}\right] = s^2 F(s) - sf(0^-) - \frac{df(t)}{dt}\bigg|_{t=0^-}. \tag{3.13}$$

To understand the value of this property, consider the following linear second-order ordinary differential equation[1]:

$$\frac{d^2 f(t)}{dt^2} + \frac{df(t)}{dt} = t^2 \tag{4.1}$$

with initial conditions $f(0^-) = 0$ and $\frac{df(t)}{dt}\bigg|_{t=0^-} = 0$.

Before trying to solve an equation like this, it is often helpful to think about what the equation is telling you – that is, what does each term of this

---

[1] This equation is "linear" because the each of the derivatives is raised to the first power, "second-order" because the highest-order derivative is a second derivative, "ordinary" because the function $f(t)$ depends on a single variable $(t)$, and "differential" because it involves derivatives of the function $f(t)$.

equation mean? In this case, recall that the first derivative $\frac{df(t)}{dt}$ represents the slope of the function $f(t)$ plotted over time, and the second derivative $\frac{d^2 f(t)}{dt^2}$ is the change in that slope over time, which is a measure of the curvature of the function. So Eq. 4.1 says that the sum of the curvature and the slope of the function $f(t)$ over time equals the square of the value of time, and the initial conditions tell you that both the function and its slope have a value of zero at time $t = 0^-$ (the instant before time $t = 0$, as described in Chapter 1).

You may know a way (or several ways) of solving an equation like this, but consider what happens when you take the Laplace transform of both sides of Eq. 4.1:

$$\mathcal{L}\left[\frac{d^2 f(t)}{dt} + \frac{df(t)}{dt}\right] = \mathcal{L}\left[t^2\right]. \tag{4.2}$$

As described in Section 3.1, the linearity property of the Laplace transform means that you can apply the transform to each term on the left side individually:

$$\mathcal{L}\left[\frac{d^2 f(t)}{dt}\right] + \mathcal{L}\left[\frac{df(t)}{dt}\right] = \mathcal{L}\left[t^2\right]. \tag{4.3}$$

Now apply the second-order derivative property of the Laplace transform (Eq. 3.13) to the first term and the first-order derivative property (Eq. 3.12) to the second term. That results in an algebraic equation:

$$s^2 F(s) - sf(0^-) - \left.\frac{df}{dt}\right|_{t=0^-} + sF(s) - f(0^-) = \mathcal{L}\left[t^2\right], \tag{4.4}$$

and applying the initial conditions given above makes this

$$s^2 F(s) + sF(s) = \mathcal{L}\left[t^2\right]. \tag{4.5}$$

For the right side of this equation, you can use the result shown in Section 2.4 of Chapter 2:

$$\mathcal{L}\left[t^n\right] = \frac{n!}{s^{(n+1)}}, \tag{4.6}$$

which for $n = 2$ is

$$\mathcal{L}\left[t^2\right] = \frac{2!}{s^{(2+1)}} = \frac{2}{s^3}. \tag{4.7}$$

So taking the Laplace transform of both sides of Eq. 4.1 leads to the following equation:

$$s^2 F(s) + sF(s) = \frac{2}{s^3} \tag{4.8}$$

or

$$s(s + 1)F(s) = \frac{2}{s^3}. \tag{4.9}$$

Hence

$$F(s) = \frac{2}{s^3(s)(s + 1)} = \frac{2}{s^4(s + 1)}. \tag{4.10}$$

This is the Laplace transform of the solution to the differential equation, so taking the inverse Laplace transform of the s-domain function $F(s)$ will give the time-domain solution $f(t)$. The most straightforward way to find that inverse Laplace transform is to use partial-fraction decomposition to write Eq. 4.10 as

$$F(s) = \frac{2}{s^4(s + 1)} = 2\left[\frac{1}{s^4} - \frac{1}{s^3} + \frac{1}{s^2} - \frac{1}{s} + \frac{1}{s + 1}\right] \tag{4.11}$$

(if you need help getting this result, see the problems at the end of this chapter and the online solutions, and remember that there is an overview of partial fractions with links to helpful resources on this book's website).

With $F(s)$ in this form, $f(t)$ can be written as the inverse transform:

$$f(t) = \mathcal{L}^{-1}\left[2\left(\frac{1}{s^4} - \frac{1}{s^3} + \frac{1}{s^2} - \frac{1}{s} + \frac{1}{s + 1}\right)\right] \tag{4.12}$$

and, since the Laplace transform and its inverse are linear processes, this is

$$f(t) = 2\left[\mathcal{L}^{-1}\left(\frac{1}{s^4}\right) - \mathcal{L}^{-1}\left(\frac{1}{s^3}\right) + \mathcal{L}^{-1}\left(\frac{1}{s^2}\right)\right.$$
$$\left. -\mathcal{L}^{-1}\left(\frac{1}{s}\right) + \mathcal{L}^{-1}\left(\frac{1}{s + 1}\right)\right].$$

With $f(t)$ in this form, you can use the inverse of Eq. 4.6 to find the inverse Laplace transform of the first four terms on the right side:

$$\mathcal{L}^{-1}\left[\frac{n!}{s^{(n+1)}}\right] = t^n \tag{4.13}$$

or

$$\mathcal{L}^{-1}\left[\frac{1}{s^{(n+1)}}\right] = \frac{1}{n!}t^n. \tag{4.14}$$

Use this equation with $n = 3$ for the first term on the right side of Eq. 4.12, with $n = 2$ for the second term, $n = 1$ for the third term, and $n = 0$ for the fourth term.

That leaves only the rightmost term of Eq. 4.12 with $s + 1$ in the denominator, and, to find the inverse Laplace transform of that term, recall from Section 2.2 that $\mathcal{L}[e^{at}] = \frac{1}{s-a}$ in the region of convergence of $s > a$, so

$$\mathcal{L}^{-1}\left[\frac{1}{s-a}\right] = e^{at}. \tag{4.15}$$

This can be applied to the rightmost term of Eq. 4.12 by setting $a = -1$.

Putting the five terms together gives $f(t)$ as

$$f(t) = 2\left[\frac{1}{3!}t^3 - \frac{1}{2!}t^2 + \frac{1}{1!}t^1 - \frac{1}{0!}t^0 + e^{(-1)t}\right]$$

or

$$f(t) = \frac{1}{3}t^3 - t^2 + 2t - 2 + 2e^{-t}. \tag{4.16}$$

You can verify that this is a solution to the differential equation Eq. 4.1 by taking the first and second temporal derivatives of $f(t)$ and confirming that their sum is $t^2$.

In this case, the coefficients of the derivatives are constant and the initial conditions are both zero, but the approach of using the Laplace transform to convert a differential equation into an algebraic equation works even when the coefficients are variable and the initial conditions are nonzero. You can see some examples of that in the other sections of this chapter and in the chapter-end problems and online solutions.

Although many problems in physics and engineering involve ordinary differential equations such as that discussed above, many more situations are described by functions of more than one variable. In that case, the relevant equation may be a partial differential equation, and the Laplace transform can be helpful in solving that type of equation, as well.

To understand how that works, consider a function that depends not only on time but on space as well. For example, the function $f(x,t)$ depends on time ($t$) and on a single spatial dimension ($x$). In that case, the derivative property of the Laplace transform looks like this:

$$\mathcal{L}\left[\frac{\partial f(x,t)}{\partial t}\right] = sF(x,s) - f(x,0^-), \tag{4.17}$$

which is the multivariable version of Eq. 3.12; note that the derivative is taken with respect to the time variable.

As you might expect, this can be extended to higher-order time derivatives using a modified version of Eq. 3.15. For second-order time derivatives, it looks like this:

$$\mathcal{L}\left[\frac{\partial^2 f(x,t)}{\partial t^2}\right] = s^2 F(x,s) - sf(x,0^-) - \left.\frac{\partial f(x,t)}{\partial t}\right|_{t=0^-}. \tag{4.18}$$

Converting time derivatives into multiplication by a power of $s$ and subtracting initial values is useful in many situations, but you may be wondering "What about differential equations that involve the partial derivative of $f(x,t)$ with respect to $x$ rather than (or in addition to) the partial derivative with respect to $t$?" In such cases, the Laplace transform may also be helpful by converting a partial differential equation into an ordinary differential equation.

To see how that works, note that the integral in the Laplace transform is taken over the time variable, and Leibnitz's rule says that

$$\int_a^b \frac{\partial f(x,t)}{\partial x}dt = \frac{d}{dx}\left(\int_a^b f(x,t)dt\right) \tag{4.19}$$

as long as the integration limits are constant. Note that the derivative on the right side of this equation is written as an ordinary rather than as a partial derivative, since the time dependence of $f(x,t)$ is "integrated out" by the integration process. Thus

$$\mathcal{L}\left[\frac{\partial f(x,t)}{\partial x}\right] = \int_{0^-}^\infty \frac{\partial f(x,t)}{\partial x}e^{-st}dt = \frac{d}{dx}\left(\int_{0^-}^\infty f(x,t)e^{-st}dt\right)$$

$$= \frac{d}{dx}F(x,s). \tag{4.20}$$

Hence the partial derivative of the space- and time-domain function $f(x,t)$ with respect to $x$ can be pulled outside the transform integral, becoming an ordinary derivative of $F(x,s)$ for which the time dependence has been integrated out.

And if the differential equation involves a second-order partial derivative with respect to $x$, the same approach can be used:

$$\mathcal{L}\left[\frac{\partial^2 f(x,t)}{\partial x^2}\right] = \frac{d^2}{dx^2}[F(x,s)]. \tag{4.21}$$

The concepts and mathematical techniques discussed in this section can be applied to a variety of differential equations that arise in specific applications. You can see several of those applications in the other sections of this chapter and in the chapter-end problems.

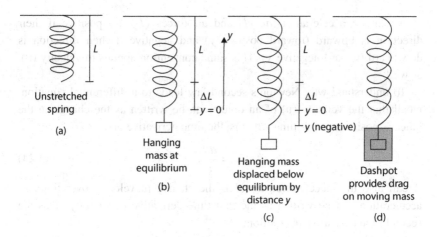

Figure 4.1 Parameters of a hanging mass oscillating in one dimension.

## 4.2 Mechanical Oscillations

An instructive application of the Laplace transform comes about by considering the situation of a mass hanging from a spring as shown in Fig. 4.1. Considering only vertical motion (that is, ignoring any side-to-side movement) makes this a one-dimensional problem, so if you define the $y$-axis as pointing vertically upward, the objective is to determine the position $y(t)$ of the mass as a function of time after some initial displacement.

As is often the case in mechanics problems, the position of an object as a function of time can be determined by solving the "equation of motion" relevant to the situation. In this case, the equation of motion can be derived from Newton's second law relating the vector acceleration[2] $\overrightarrow{accel}$ of an object of mass $m$ to the vector sum of the forces $\Sigma \vec{F}$ acting on the object:

$$\overrightarrow{accel} = \frac{\Sigma \vec{F}}{m}. \tag{4.22}$$

For one-dimensional motion along the $y$-axis, this vector equation can be written using the scalar quantities $accel$ for acceleration and $\Sigma F$ for the sum of the forces as long as the directional information is included by using the appropriate sign:

$$accel = \frac{\Sigma F}{m}, \tag{4.23}$$

[2] In this section, acceleration is represented by "$accel$" because the single letter "$a$" will be used for a different variable.

in which the acceleration (*accel*) and all forces (*F*) are positive if their direction is upward (toward positive $y$) and negative if their direction is downward (toward negative $y$). This same convention applies to velocity ($v$), as well.

To understand why Newton's second law leads to a differential equation, recall that the velocity ($v$) of an object can be written as the change in the object's position $y$ over time – that is, the time derivative of $y$:

$$v = \frac{dy}{dt}. \qquad (4.24)$$

Recall also that acceleration *accel* is the change in velocity over time, so acceleration can be written as the first time derivative of velocity or as the second time derivative of position:

$$accel = \frac{dv}{dt} = \frac{d^2y}{dt^2}. \qquad (4.25)$$

Using this expression for acceleration makes Newton's second law (Eq: 4.23) look like this:

$$\frac{d^2y}{dt^2} = \frac{1}{m}\Sigma F, \qquad (4.26)$$

which is a linear second-order differential equation that is the basis for the equation of motion of the hanging mass.

Before solving that equation, it is necessary to determine each of the forces acting on the object and then to express those forces in terms of known quantities. As long as the system is near the surface of the Earth, one force you must consider is the force of gravity $F_g$:

$$F_g = -mg, \qquad (4.27)$$

in which $m$ represents the mass (SI units of kg) and $g$ is the acceleration due to gravity (SI units of m/sec$^2$). At the Earth's surface, $g$ has a magnitude of 9.8 m/sec$^2$, and the minus sign in this equation encapsulates the fact that the force of gravity operates downward, which is defined as the negative-$y$ direction in this case.

The spring also exerts a force on the hanging mass, and to find an expression for the spring force $F_s$, begin by calling the unstretched length of the spring, $L$, as shown in Fig. 4.1a. Hanging mass $m$ on the spring causes it to stretch, and Hooke's law tells you that the force of the spring increases linearly with distance from the unstretched length:

$$F_s = -ky_d, \qquad (4.28)$$

in which $k$ represents the spring constant (with SI units of Newtons per meter [N/m]), $y_d$ represents the displacement of the bottom end of the spring from the unstretched position (with SI units of meters), and the minus sign indicates that the spring force is always in the opposite direction from the displacement. So with the $y$-axis pointing vertically upward, $y_d$ is negative as the spring stretches, which means that the spring force is upward (toward positive $y$) in that case.

When the upward force of the stretched spring exactly balances the downward force of gravity, the mass is in the equilibrium position, which you can define as $y = 0$, as shown in Fig. 4.1b . At the equilibrium position, the spring has stretched by amount $\Delta L$ (taken as positive for a stretched spring), so Eq. 4.28 tells you that the upward force of the spring is $-ky_d = k\Delta L$ at that point (remember that $y_d$ is negative while $\Delta L$ is positive).

Knowing that the spring force at equilibrium equals $k\Delta L$ is useful, because you know that at equilibrium the acceleration is zero – the mass will stay put at this position if its initial velocity is zero. And zero acceleration means that the total force acting on the mass must be zero:

$$accel = \frac{\Sigma F}{m} = 0. \tag{4.29}$$

So if the only forces acting on the mass are the force of gravity ($F_g = -mg$) and the spring force ($F_s = k\Delta L$ at equilibrium), then

$$\Sigma F = F_g + F_s = 0,$$

and

$$F_s = -F_g,$$
$$k\Delta L = -(-mg) = mg. \tag{4.30}$$

Now consider what happens if a temporary external force (such as your hand) pulls the mass down below the equilibrium position, as shown in Fig. 4.1c. If the new position of the mass (that is, the attachment point of the mass to the bottom of the spring) is $y$ (a negative number), the upward spring force increases by the amount $-ky$, while the downward force of gravity remains $-mg$. That makes Newton's second law look like this:

$$accel = \frac{\Sigma F}{m} = \frac{1}{m}(F_g + F_s)$$

$$= \frac{1}{m}(-mg + k\Delta L - ky)$$

$$= \frac{1}{m}(-mg + mg - ky) = -\left(\frac{k}{m}\right)y. \tag{4.31}$$

If the temporary force that pulled the mass to a position below equilibrium is then released, the upward force of the spring will cause the mass to accelerate upward. When the mass reaches the equilibrium point, the total force acting on it is zero, so the mass passes through equilibrium and continues moving upward. Above equilibrium the upward spring force is less than the downward force of gravity, and the net downward force causes the upward speed of the mass to decrease until the mass stops and begins to accelerate downward. After the mass passes downward through the equilibrium point, the process repeats. This process is called "simple harmonic motion," and writing the acceleration as the second derivative of the position with respect to time turns Eq. 4.31 into the differential equation of this motion:

$$accel = \frac{d^2 y}{dt^2} = -\left(\frac{k}{m}\right) y$$

or

$$\frac{d^2 y}{dt^2} + \left(\frac{k}{m}\right) y = 0. \tag{4.32}$$

Before using the Laplace transform to find the solution to this equation, consider the effect of adding another force to the situation, such as a drag force that opposes the motion of the hanging mass. Such a force arises if the mass is immersed in a fluid, as shown in Fig. 4.1d. In this arrangement, sometimes called a "dashpot," the magnitude of the drag force is directly proportional to the speed of the mass, and the direction of the drag force is always opposite to the direction in which the mass is moving.

The constant of proportionality between the speed of the mass ($v = dy/dt$) and the drag force ($F_d$) is the coefficient of drag ($c_d$), so in the one-dimensional case the drag force may be written as

$$F_d = -c_d v = -c_d \frac{dy}{dt}, \tag{4.33}$$

in which the SI units of the drag coefficient $c_d$ are N/(m/s) and the minus sign accounts for the fact that the drag force always points in the opposite direction from the velocity.

Adding the drag force to the force of gravity and the spring force makes the one-dimensional version of Newton's second law look like this:

$$accel = \frac{\Sigma F}{m} = \frac{1}{m}(F_g + F_s + F_d), \tag{4.34}$$

and the equation of motion (Eq. 4.32) becomes

$$\frac{d^2y}{dt^2} = \frac{1}{m}\left(-ky - c_d\frac{dy}{dt}\right)$$

or

$$\frac{d^2y}{dt^2} + \left(\frac{c_d}{m}\right)\frac{dy}{dt} + \left(\frac{k}{m}\right)y = 0. \qquad (4.35)$$

So with the gravitational force, spring force, and drag force acting on the hanging mass, the equation of motion for the time-dependent position $y(t)$ is an ordinary second-order differential equation containing one term with the second time derivative of $y(t)$, one term with the first time derivative of $y(t)$, and one term with $y(t)$.

One approach to finding $y(t)$ is to use the Laplace transform to convert this differential equation into an algebraic equation. In the notation of earlier chapters, the position function $y(t)$ serves as the time-domain function $f(t)$, and the Laplace transform $\mathcal{L}[y(t)] = Y(s)$ is the $s$-domain function $F(s)$. Taking the Laplace transform of both sides of Eq. 4.35 gives

$$\mathcal{L}\left[\frac{d^2y}{dt^2} + \left(\frac{c_d}{m}\right)\frac{dy}{dt} + \left(\frac{k}{m}\right)y\right] = \mathcal{L}[0], \qquad (4.36)$$

and the linearity property of the Laplace transform means that you can apply the transform to each term individually and that constant terms can be pulled outside the transform:

$$\mathcal{L}\left[\frac{d^2y}{dt^2}\right] + \left(\frac{c_d}{m}\right)\mathcal{L}\left[\frac{dy}{dt}\right] + \left(\frac{k}{m}\right)\mathcal{L}[y] = \mathcal{L}[0]. \qquad (4.37)$$

With the equation of motion in this form, the time-derivative properties of the Laplace transform discussed in Section 3.4 can be applied. For the second-order time derivative in the first term in the equation of motion, Eq. 3.13 gives:

$$\mathcal{L}\left[\frac{d^2y}{dt^2}\right] = s^2Y(s) - sy(0^-) - \frac{dy}{dt}\bigg|_{t=0^-}, \qquad (4.38)$$

and for the first-order derivative in the second term in the equation of motion, Eq. 3.12) gives

$$\mathcal{L}\left[\frac{dy}{dt}\right] = sY(s) - y(0^-). \qquad (4.39)$$

Now use the definition of the Laplace transform on the remaining term:

$$\mathcal{L}[y] = Y(s), \tag{4.40}$$

and, finally, use the fact that the Laplace transform of 0 is also 0:

$$\mathcal{L}[0] = 0. \tag{4.41}$$

Thus the algebraic version of the equation of motion for the hanging mass is

$$s^2 Y(s) - s y(0) - \frac{dy}{dt}\Big|_{t=0} + \left(\frac{c_d}{m}\right)[sY(s) - y(0)] + \left(\frac{k}{m}\right) Y(s) = 0. \tag{4.42}$$

This equation makes it clear that the $s$-domain function $Y(s)$, and therefore the time-domain function $y(t)$, depend on the initial conditions, specifically on the starting position $y(0)$ and the initial speed $dy/dt$ at time $t = 0$. If the starting position is taken as $y_0$ and the speed is zero at time $t = 0$, the equation of motion becomes

$$s^2 Y(s) - s y_0 - 0 + \left(\frac{c_d}{m}\right)[sY(s) - y_0] + \left(\frac{k}{m}\right) Y(s) = 0, \tag{4.43}$$

which can be rearranged to give

$$Y(s)\left[s^2 + \left(\frac{c_d}{m}\right)s + \left(\frac{k}{m}\right)\right] = s y_0 + \left(\frac{c_d}{m}\right) y_0 \tag{4.44}$$

or

$$Y(s) = \frac{s y_0 + \left(\frac{c_d}{m}\right) y_0}{s^2 + \left(\frac{c_d}{m}\right)s + \left(\frac{k}{m}\right)}. \tag{4.45}$$

This is the $s$-domain solution to the differential equation of motion of the hanging mass, but if you want to know the position $y(t)$ of the mass as a function of time you must take the inverse Laplace transform of $Y(s)$. For many problems (including this one), the most straightforward approach to finding the inverse Laplace transform is to attempt to convert the $s$-domain function into a form recognizable as the Laplace transform of a known function, such as the basic functions used in the examples in Chapter 2.

In this case, one way to do that is to begin by completing the square in the denominator of the expression for $Y(s)$ (Eq. 4.45):

$$s^2 + \left(\frac{c_d}{m}\right)s + \left(\frac{k}{m}\right) = \left(s + \frac{c_d}{2m}\right)^2 - \left(\frac{c_d}{2m}\right)^2 + \left(\frac{k}{m}\right).$$

To make this recognizable as one of the basic functions from the examples in Chapter 2, start by defining a variable $a$ (not the acceleration) as

$$a = \frac{c_d}{2m} \tag{4.46}$$

and note that the units of $a$ are the units of the drag coefficient $c_d$ divided by the units of $m$. Since the SI units of $c_d$ are N/(m/sec) and the SI units of $m$ are kg, the SI units of $a$ are N/(m/sec)/kg. But the units N are the same as kg m/sec$^2$, so the SI units of $a$ are kg(m/sec$^2$)/(m/sec)/kg), which reduces to 1/sec. Thus $a$ has units of frequency, and as you'll see below, this variable plays a role in $y(t)$ not of an oscillation frequency (cycles/sec), but as an amplitude-variation frequency (number of $1/e$ steps per second).

With this definition of $a$, the denominator becomes

$$s^2 + \left(\frac{c_d}{m}\right)s + \left(\frac{k}{m}\right) = (s + a)^2 - a^2 + \left(\frac{k}{m}\right).$$

Now define an angular frequency related to the spring constant and the mass of the hanging object:

$$\omega_0^2 = \frac{k}{m}; \tag{4.47}$$

the reason for defining this angular frequency will become clear when you have an expression for $y(t)$. Note that the SI units of $\omega_0^2$ are (N/m)/kg, which reduce to 1/sec$^2$, so $\omega_0$ has units of 1/sec (and remember that radians are dimensionless, so this is equivalent to rad/sec).

Using this definition makes the denominator look a bit more familiar if you have worked through the examples in Chapter 2:

$$s^2 + \left(\frac{c_d}{m}\right)s + \left(\frac{k}{m}\right) = (s + a)^2 - a^2 + \omega_0^2;$$

and one more definition should make the denominator even more familiar:

$$\omega_1^2 = \omega_0^2 - a^2. \tag{4.48}$$

This definition of $\omega_1$ is useful long as $\omega_0^2$ is greater than $a^2$; cases in which $\omega_0^2$ is equal to or smaller than $a^2$ are discussed below.

The denominator of Eq. 4.45 now looks like this:

$$s^2 + \left(\frac{c_d}{m}\right)s + \left(\frac{k}{m}\right) = (s + a)^2 + \omega_1^2$$

so

$$Y(s) = \frac{s y_0 + \left(\frac{c_d}{m}\right) y_0}{(s + a)^2 + \omega_1^2}. \tag{4.49}$$

Now consider the numerator, which can be written as

$$s y_0 + \left(\frac{c_d}{m}\right) y_0 = s y_0 + 2 a y_0 = s y_0 + a y_0 + a y_0$$
$$= (s + a) y_0 + a y_0$$

and plugging this in for the numerator of Eq. 4.49 makes it

$$Y(s) = \frac{(s + a) y_0 + a y_0}{(s + a)^2 + \omega_1^2} = y_0 \frac{(s + a)}{(s + a)^2 + \omega_1^2} + y_0 \frac{a}{(s + a)^2 + \omega_1^2}. \tag{4.50}$$

This is the complex-frequency domain function that satisfies the partial differential equation representing the equation of motion of the hanging mass. As stated above, if you want to know the position $y(t)$ of the mass as a function of time, you need to take the inverse Laplace transform of $Y(s)$.

To do that, look carefully at each of the terms in Eq. 4.50. If the first term looks familiar, you may be remembering the function $F(s)$ resulting from the Laplace transform of the time-domain function $\cos(\omega_1 t)$. Of course, this function includes $s + a$ rather than $s$, but recall the shift property of the Laplace transform discussed in Section 3.2. That property says that replacing $s$ by $s + a$ in the complex-frequency domain function $F(s)$ has the effect of multiplying the corresponding time-domain function $f(t)$ by the factor $e^{-at}$. So the inverse Laplace transform of the first term of Eq. 4.50 is

$$\mathcal{L}^{-1}\left[y_0 \frac{(s + a)}{(s + a)^2 + \omega_1^2}\right] = y_0 e^{-at} \cos(\omega_1 t). \tag{4.51}$$

Now consider the second term of Eq. 4.50. Multiplying this term by $\frac{\omega_1}{\omega_1}$ gives

$$y_0 \frac{a}{(s + a)^2 + \omega_1^2} = \left(\frac{a y_0}{\omega_1}\right) \frac{\omega_1}{(s + a)^2 + \omega_1^2}, \tag{4.52}$$

which has the form of the function $F(s)$ resulting from the Laplace transform of the time-domain function $\sin(\omega_1 t)$. So the inverse Laplace transform of the second term is

$$\mathcal{L}^{-1}\left[\left(\frac{a y_0}{\omega_1}\right) \frac{\omega_1}{(s + a)^2 + \omega_1^2}\right] = \left(\frac{a y_0}{\omega_1}\right) e^{-at} \sin(\omega_1 t). \tag{4.53}$$

Putting the terms together gives the time-domain function $y(t)$ that is the solution to the equation of motion for the hanging mass with $\omega_0^2 > a^2$:

$$y(t) = y_0 e^{-at} \cos(\omega_1 t) + \left(\frac{a y_0}{\omega_1}\right) e^{-at} \sin(\omega_1 t). \tag{4.54}$$

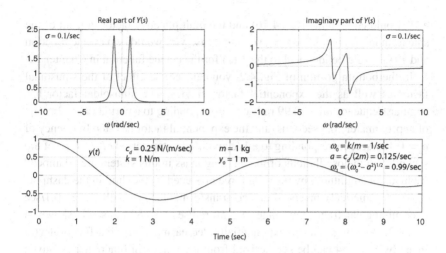

Figure 4.2 Example solutions $Y(s)$ and $y(t)$ for damped hanging mass.

Before considering the meaning of this expression, it is worthwhile to make sure that $y(t)$ satisfies the equation of motion (Eq. 4.35) as well as the boundary conditions. Taking the first and second time derivatives and doing the appropriate summation of the results shows that Eq. 4.54 does indeed satisfy the equation of motion and the initial conditions of position $y(0) = y_d$ and velocity $(dy/dt = 0)$ at time $t = 0$ (if you need help proving that, see the chapter-end problems and online solutions).

Once you are confident that Eq. 4.54 meets the requirements for the solution to this problem, you may find it helpful to look at the terms of $y(t)$ to see what they are telling you about the behavior of the system over time. Both terms have a sinusoidal factor that oscillates between $+1$ and $-1$ with angular frequency of $\omega_1$, and the amplitude of those oscillations decreases exponentially over time due to the factor of $e^{-at}$. Additionally, the amplitude of the cosine term is weighted by a factor of $y_0$ (the initial displacement), and the sine term has additional amplitude weighting $a/\omega_1$, which is the ratio of the exponential-decay frequency to the oscillation frequency.

The real and imaginary parts of the resulting $s$-domain function $Y(s)$ and the time-domain function $y(t)$ are shown in Fig. 4.2 with parameters $c_d = 0.25$ N/(m/sec), $k = 1$ N/m, $m = 1$ kg, and $y_0 = 1$ m. This combination of parameters makes the cosine term of $y(t)$ somewhat larger than the sine term, so it shouldn't be too surprising that the $s$-domain function $Y(s)$ resembles the Laplace transform derived for the cosine function in Section 2.3, such as that shown in Fig. 2.13. In the limit of zero drag ($c_d = 0$), the exponential factor

$e^{-at}$ would equal unity (Eq. 4.46) and the multiplicative factor in front of the sine term in $y(t)$ would be zero, so in that case $y(t)$ would be a cosine function and $Y(s)$ would be identical to the $F(s)$ for the cosine function in Chapter 2.

In the bottom portion of Fig. 4.2, you can see the effect of the sinusoidal factor as well as the exponential factor in $y(t)$. The sinusoidal factor has angular frequency $\omega_1 = 0.99$ rad/sec, corresponding to a period $(T = 2\pi/\omega_1)$ of approximately 6.3 seconds, and the exponential factor has step frequency of $a = 0.125$/sec, corresponding to a $1/e$ amplitude-step period $(T = 1/a)$ of 8.0 seconds. So the oscillations of the hanging mass in this system are "damped" (reduced in amplitude) by the drag force produced by the fluid in the dashpot.

The (numerical) inverse Laplace transform of $Y(s)$ with $\sigma = 0.1$/sec over an angular frequency range of $-50$ rad/sec to $+50$ rad/sec is shown in Fig. 4.3, verifying that the exponentially decreasing sinusoidal function $y(t)$ given by Eq. 4.54 can be synthesized from the $s$-domain function $Y(s)$ given by Eq. 4.50.

Of course, the behavior shown in Fig. 4.3 depends on the values of the spring constant, mass, drag coefficient, and initial displacement. And as mentioned above, the time-domain function $y(t)$ given by Eq. 4.54 pertains to situations in which $\omega_0^2$ is greater than $a^2$.

To determine the behavior of the system when that is not the case (that is, when $a^2$ is greater than $\omega_0^2$), start by defining $\omega_1$ as

$$\omega_1^2 = a^2 - \omega_0^2. \tag{4.55}$$

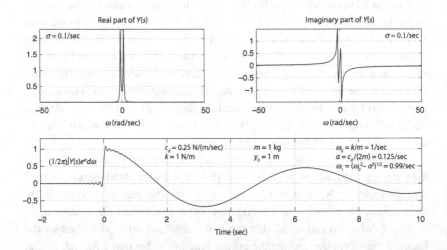

Figure 4.3   $y(t)$ synthesized by numerically integrating $Y(s)e^{st}$.

With this definition, the $s$-domain function $Y(s)$ becomes

$$Y(s) = y_0 \frac{(s+a)}{(s+a)^2 - \omega_1^2} + y_0 \frac{a}{(s+a)^2 - \omega_1^2}. \qquad (4.56)$$

Just as in the previous case, the next step is to find the inverse Laplace transform of $Y(s)$. The inverse transform of the first term is

$$\mathcal{L}^{-1}\left[ y_0 \frac{(s+a)}{(s+a)^2 - \omega_1^2} \right] = y_0 e^{-at} \cosh(\omega_1 t) \qquad (4.57)$$

and the inverse Laplace transform of the second term (again multiplied by $\omega_1/\omega_1$) is

$$\mathcal{L}^{-1}\left[ \left(\frac{a y_0}{\omega_1}\right) \frac{\omega_1}{(s+a)^2 - \omega_1^2} \right] = \left(\frac{a y_0}{\omega_1}\right) e^{-at} \sinh(\omega_1 t). \qquad (4.58)$$

Adding these terms together gives $y(t)$:

$$y(t) = y_0 e^{-at} \cosh(\omega_1 t) + \left(\frac{a y_0}{\omega_1}\right) e^{-at} \sinh(\omega_1 t), \qquad (4.59)$$

in which the sinusoidal functions in the previous case have been replaced by the hyperbolic sinusoidal functions, while the exponential factors and multiplicative constants remain the same.

You can see plots of the $s$-domain function $Y(s)$ and the time-domain function $y(t)$ for a system with $a^2 > \omega_0^2$ in Fig. 4.4. In this case the mass

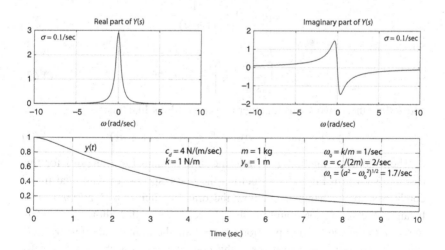

Figure 4.4 Example solutions $Y(s)$ and $y(t)$ for overdamped hanging mass.

remains 1 kg, the spring constant remains 1 N/m, and the initial conditions are the same, but the drag coefficient has been increased from 0.25 to 4 N/(m/s). Notice that this greater drag prevents the system from completing even one oscillation as the mass slowly works its way toward the equilibrium point; this system is "overdamped".

One more situation of interest occurs when $\omega_0^2 = a^2$. That means that $k = c_d^2/4m$ and $\omega_1 = 0$, and in that case the the $s$-domain function $Y(s)$ is

$$Y(s) = y_0 \frac{(s+a)}{(s+a)^2} + y_0 \frac{a}{(s+a)^2}. \tag{4.60}$$

As in the previous two cases, start by taking the inverse Laplace transform of the first term. To do that, remember from Section 2.2 that

$$\mathcal{L}[e^{-at}] = \frac{1}{s+a},$$

so the inverse Laplace transform of the first term is

$$\mathcal{L}^{-1}\left[y_0 \frac{(s+a)}{(s+a)^2}\right] = y_0 \mathcal{L}^{-1}\left[\frac{1}{s+a}\right] = y_0 e^{-at}. \tag{4.61}$$

Now take the inverse Laplace transform of the second term. Recall from Section 2.4 that

$$\mathcal{L}[t^n] = \frac{n!}{s^{n+1}} = \frac{1}{s^2} \text{ for } n = 1$$

and applying the frequency-shifting property (Section 3.2) gives

$$\mathcal{L}[e^{-at}t] = \frac{1}{(s+a)^2}.$$

So

$$\mathcal{L}^{-1}\left[\frac{ay_0}{(s+a)^2}\right] = ay_0 t e^{-at} \tag{4.62}$$

and adding these two terms together makes $y(t)$ look like this:

$$y(t) = y_0 e^{-at} + ay_0 t e^{-at} = y_0 e^{-at}(1+at). \tag{4.63}$$

The $s$-domain function $Y(s)$ and the time-domain function $y(t)$ for this case are shown in Fig. 4.5. As before, the mass, spring constant, and initial conditions are unchanged, but now the drag coefficient is 2 N/(m/s). This balances the spring force and the drag force in a way that allows the system to return to the equilibrium position (and remain there) in the shortest possible time without oscillating. Systems in this condition are described as "critically damped."

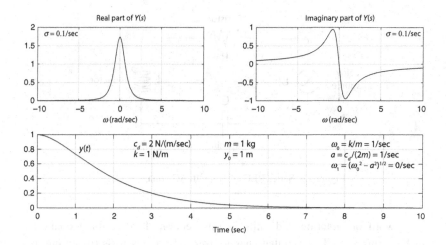

Figure 4.5 Example solutions $Y(s)$ and $y(t)$ for critically damped hanging mass.

One reason that the analysis of a mass hanging on a spring gets a lot of attention in physics and engineering classes is that the differential equation describing the position of the object is quite similar to equations pertaining to very different situations. You can see one of those situations in the next section.

## 4.3 Electric-Circuit Oscillations

In this section, you'll see how to apply the Laplace transform to the solution of the differential equation that describes the flow of charge in an electric circuit that contains a resistor, inductor, and capacitor in series. If you have studied circuits, you may remember that connecting devices "in series" means that all the current that flows through one device also flows through the other devices, so in a series RLC circuit such as that shown in Fig. 4.6 there is only a single loop, with no branch points at which some of the current may divert into a different path.

As you can see on the left side of this figure, this series RLC circuit may also contain a voltage source, and the voltage (also called the "potential difference" or "electromotive force [emf]") produced by that device may vary with time, as indicated by the little sine wave shown in the device's symbol. A voltage source with output that varies periodically is often called an "alternating current" (AC) generator, in contrast to a fixed-potential or "direct current" (DC) device such as a battery.

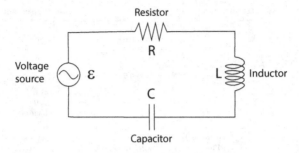

Figure 4.6 Series RLC circuit with AC voltage source.

A word on notation and units in this section: In describing quantities such as current and charge that change over time, many texts (including this one) use lower-case Roman letters, such as $i(t)$ for AC current and $q(t)$ for charge. However, fixed-device characteristics such as resistance, inductance, and capacitance are typically written using upper-case roman letters, so $R$ for resistance, $L$ for inductance, and $C$ for capacitance.

Concerning units: the base SI unit of charge is the coulomb (written as "C") and the base SI unit of current is the ampere (written as "A" and equivalent to coulombs per second), while most other circuit quantities have derived units (that is, units made up of combinations of base units). So potential difference has SI units of volts (written as V, with base units of kg m$^2$/(sec$^3$ A)), resistance has units of ohms (written as $\Omega$ and with base units of kg m$^2$/(sec$^3$ A$^2$)), inductance has units of henries (base units of kg m$^2$/(sec$^2$ A$^2$)), and capacitance has units of farads (written as 'H' and with base units of sec$^4$ A$^2$/(m$^2$ kg). These relations to base units are presented here because they may help you understand why certain combinations of quantities such as $\sqrt{1/LC}$ and $R/(2L)$ have units of frequency, and also because checking units is often a good way to verify the consistency of an equation with many terms.

So how does a differential equation describe the behavior of this type of circuit? To understand that, recall that the Kirchhoff voltage law says that the sum of the voltage changes across the elements of a series circuit must add to zero. For each element in the circuit, the change in voltage depends on the nature of the device under consideration. Imagine making an "analysis loop" around the circuit in which you keep track of the potential change across each circuit element (you can do this in either direction as long as you continue in the same direction around the entire circuit). If you make a complete loop around the circuit, you can be sure that any voltage increase across one or more of the elements will be offset by voltage decreases across other elements.

Here is what happens to the potential across each of the devices in a series RLC circuit driven by a voltage generator. Across the voltage source, the electrical potential difference between the terminals is given by the emf ($\mathcal{E}$) of the generator, which is fixed for a DC source such as a battery but which varies with time for an AC source. As you continue around your analysis loop to a resistor, if the direction in which you cross the resistor is the same as the direction in which current is flowing, you will see a decrease in potential (a voltage drop) equal to the product of the current ($i$) and the resistance ($R$). Likewise, the potential drop across the inductor is given by the product of the inductance ($L$) and the time derivative of the current ($di/dt$), and across a capacitor the potential drop is given by the ratio of the charge ($q$) to the capacitance ($C$). When considering these relationships, it is important to remember that voltage "gains" and "drops" can be positive or negative – a negative voltage gain (a decrease in potential) is a positive voltage drop, and a negative voltage drop is a positive voltage gain (an increase in potential).

But whatever the sign and magnitude of the potential change across each of the devices in the circuit, you can be sure that the voltage gain of the voltage source equals the sum of the voltage drops across the resistor, capacitor, and inductor. Thus Kirchhoff's voltage law leads to this differential equation:

$$\mathcal{E} = Ri + L\frac{di}{dt} + \frac{q}{C}. \tag{4.64}$$

Since electric current consists of moving electric charge, the relationship between the current ($i$) and the charge ($q$) on the capacitor is

$$i = \frac{dq}{dt}, \tag{4.65}$$

and the time-varying potential difference of an AC voltage source can be written as

$$\mathcal{E} = V_s \sin(\omega_s t), \tag{4.66}$$

in which $V_s$ represents the amplitude of the voltage source and $\omega_s$ is the angular frequency of the voltage source.

Inserting these expressions for AC current and emf into the voltage-drop equation (Eq. 4.64) gives

$$\mathcal{E} = V_s \sin(\omega_s t) = R\frac{dq}{dt} + L\frac{d}{dt}\left(\frac{dq}{dt}\right) + \frac{q}{C} = R\frac{dq}{dt} + L\frac{d^2q}{dt^2} + \frac{q}{C}, \tag{4.67}$$

which is a second-order ordinary differential equation amenable to solution using the Laplace transform and its inverse.

Taking the Laplace transform of both sides of this equation gives:

$$\mathcal{L}[V_s \sin(\omega_s t)] = \mathcal{L}\left[R\frac{dq}{dt} + L\frac{d^2q}{dt^2} + \frac{q}{C}\right]. \qquad (4.68)$$

The linearity property of the Laplace transform means you can apply the transform to the individual terms and that constants can be brought outside the transform operator:

$$V_s\mathcal{L}[\sin(\omega_s t)] = R\mathcal{L}\left[\frac{dq}{dt}\right] + L\mathcal{L}\left[\frac{d^2q}{dt^2}\right] + \frac{1}{C}\mathcal{L}[q]. \qquad (4.69)$$

Just as in the case of mechanical oscillations discussed in the previous section, you can use the time-derivative properties of the Laplace transform described in Section 3.4 to convert this differential equation into an algebraic equation. Using Eq. 3.15, the second-order time derivative of charge $q$ becomes

$$\mathcal{L}\left[\frac{d^2q}{dt^2}\right] = s^2 Q(s) - sq(0^-) - \left.\frac{dq}{dt}\right|_{t=0^-} \qquad (4.70)$$

and Eq. 3.12 makes the first-derivative term

$$\mathcal{L}\left[\frac{dq}{dt}\right] = sQ(s) - q(0^-). \qquad (4.71)$$

The definition of the Laplace transform can be applied to the term involving $q$:

$$\mathcal{L}[q] = Q(s). \qquad (4.72)$$

That leaves only the voltage-source term $\mathcal{E} = V_s \sin(\omega_s t)$, for which the Laplace transform is given in Section 2.3 as

$$\mathcal{L}[\mathcal{E}] = V_s\mathcal{L}[\sin(\omega_s t)] = V_s\frac{\omega_s}{s^2 + \omega_s^2}. \qquad (4.73)$$

Using these Laplace transforms of $q$ and $\mathcal{E}$, the differential equation for this series RLC circuit (Eq. 4.67) becomes the algebraic equation

$$V_s\frac{\omega_s}{s^2 + \omega_s^2} = R\left[sQ(s) - q(0^-)\right]$$

$$+ L\left[s^2 Q(s) - sq(0^-) - \left.\frac{dq}{dt}\right|_{t=0^-}\right] + \frac{1}{C}Q(s). \qquad (4.74)$$

This equation can be solved for the $s$-domain function $Q(s)$ if the initial conditions $q(0^-)$ and $dq/dt|_{t=0^-}$ are known, and the inverse Laplace transform can then be used to find the time-domain function $q(t)$.

One straightforward case has no voltage generator, just a switch that opens or closes the circuit loop through the resistor, inductor, and capacitor. In this case, the capacitor has an initial charge of $q_0$, and the switch is closed (completing the circuit and allowing current to flow) at time $t = 0$. Inserting the values $V_0 = 0$, $q(0^-) = q_0$ and $dq/dt|_{t=0^-} = 0$ into Eq. 4.74 gives

$$0 = Rs\,Q(s) - Rq_0 + Ls^2 Q(s) - Lsq_0 + \frac{1}{C}Q(s).$$

Now solve for $Q(s)$:

$$Q(s)\left(Ls^2 + Rs + \frac{1}{C}\right) = (R + Ls)q_0,$$

$$Q(s) = \frac{(R + Ls)q_0}{Ls^2 + Rs + \frac{1}{C}}.$$

To put this expression for $Q(s)$ into a form that is recognizable as the Laplace transform of a basic function, start by pulling a factor of the inductance $L$ from the denominator:

$$Q(s) = \frac{(R + Ls)q_0}{L\left(s^2 + \frac{R}{L}s + \frac{1}{LC}\right)} = \frac{\left(s + \frac{R}{L}\right)q_0}{s^2 + \frac{R}{L}s + \frac{1}{LC}} \tag{4.75}$$

and then complete the square in the denominator:

$$s^2 + \frac{R}{L}s + \frac{1}{LC} = s^2 + \frac{R}{L}s + \left(\frac{R}{2L}\right)^2 + \frac{1}{LC} - \left(\frac{R}{2L}\right)^2$$

$$= \left(s + \frac{R}{2L}\right)^2 + \frac{1}{LC} - \left(\frac{R}{2L}\right)^2.$$

Inserting this expression into the denominator of Eq. 4.75 gives

$$Q(s) = \frac{\left(s + \frac{R}{L}\right)q_0}{\left(s + \frac{R}{2L}\right)^2 + \frac{1}{LC} - \left(\frac{R}{2L}\right)^2}. \tag{4.76}$$

As in the previous section, the next step is to define several new variables, starting with $a$ (again, not the acceleration):

$$a = \frac{R}{2L}, \tag{4.77}$$

which has dimensions of resistance divided by inductance (SI units of ohms/henries), which reduce to 1/sec. So this variable $a$ also has units of frequency, serving as an amplitude-variation frequency (number of $1/e$ steps per second) in $q(t)$.

Two additional definitions are useful in describing the behavior of an RLC circuit. The first is

$$\omega_0^2 = \frac{1}{LC}, \tag{4.78}$$

which is often called the "natural angular frequency" of the circuit. In the limit of zero resistance ($R = 0$), this is the angular frequency at which an initial electric charge $q_0$ will oscillate from one plate of the capacitor to the other, flowing through the inductor along the way. With no resistance to dissipate energy as heat, this process will continue forever, with energy alternating between the electric field of the capacitor (when charge is present) and the magnetic field of the inductor (when current is flowing). This is a direct analog to an oscillating mass hanging on a spring with zero drag; in that case the energy alternates between the kinetic energy of motion and the potential energy of the spring and gravitational field.

The second useful definition is

$$\omega_1^2 = \frac{1}{LC} - \left(\frac{R}{2L}\right)^2 = \omega_0^2 - a^2, \tag{4.79}$$

often called the "damped angular frequency" of the circuit since it pertains to the case in which a resistor removes energy from the system, having an effect similar to that of drag in a mechanical system. You may see $\omega_1$ referred to as a pseudo-frequency, since frequency is technically defined as a characteristic of periodic (i.e. repeating) systems, and the changing oscillation amplitude means that the system behavior does not necessarily repeat.

Another caveat that you may have expected if you have worked through the previous section on mechanical oscillations is that this definition of $\omega_1$ is useful as long as $\omega_0^2$ is greater than $a^2$, which means $\frac{1}{LC} > \left(\frac{R}{2L}\right)^2$. Cases in which $\omega_0^2$ is equal to or smaller than $a^2$ are discussed below.

Using these definitions of $a$, $\omega_0$, and $\omega_1$, Eq. 4.76 becomes

$$Q(s) = \frac{(s + 2a)q_0}{(s + a)^2 + \omega_1^2} = q_0 \frac{(s + a) + a}{(s + a)^2 + \omega_1^2}, \tag{4.80}$$

which has the same form as the $s$-domain function $Y(s)$ (Eq. 4.50) for the hanging mass discussed in the previous section. Just as in that case, the inverse Laplace transform of this $s$-domain function has decreasing exponential ($e^{-at}$) as well as sinusoidal terms with angular frequency $\omega_1$. The resulting time-domain function is

$$q(t) = q_0 e^{-at} \cos(\omega_1 t) + q_0 \left[\frac{a}{\omega_1}\right] e^{-at} \sin(\omega_1 t), \tag{4.81}$$

which is identical to $y(t)$ for the hanging mass with the variable changes $c_d \leftrightarrow R$, $m \leftrightarrow L$, and $k \leftrightarrow \frac{1}{C}$. So the oscillations of charge in this series RLC circuit behave like the oscillations of a hanging mass under equivalent initial conditions.

Using these definitions and results, the Laplace-transform analysis of a series RLC circuit driven by a variable-voltage generator with amplitude $V_s$ and angular frequency $\omega_s$ is reasonably straightforward. Consider the case in which the initial charge on the capacitor and the initial current are both zero, so $q(0^-) = 0$ and $i(0^-) = dq/dt|_{t=0^-} = 0$. These initial conditions make Eq. 4.74

$$V_s \frac{\omega_s}{s^2 + \omega_s^2} = R[sQ(s)] + L[s^2 Q(s)] + \frac{Q(s)}{C}, \tag{4.82}$$

which can be rearranged to give

$$Q(s)\left(Ls^2 + Rs + \frac{1}{C}\right) = \frac{V_s \omega_s}{s^2 + \omega_s^2} \tag{4.83}$$

or

$$Q(s) = \frac{V_s \omega_s}{(s^2 + \omega_s^2)\left[Ls^2 + Rs + \frac{1}{C}\right]}. \tag{4.84}$$

Once again, it is helpful to pull out a factor of $L$ and to complete the square on the right term in the denominator:

$$Q(s) = \frac{V_s \omega_s / L}{(s^2 + \omega_s^2)\left[\left(s + \frac{R}{2L}\right)^2 + \frac{1}{LC} - \left(\frac{R}{2L}\right)^2\right]}, \tag{4.85}$$

and inserting the definitions for $a$ and $\omega_1$ makes this

$$Q(s) = \frac{V_s \omega_s / L}{(s^2 + \omega_s^2)[(s + a)^2 + \omega_1^2]}. \tag{4.86}$$

If you compare this expression for $Q(s)$ to Eq. 4.80 for the undriven case, you will see that the addition of an AC voltage generator has resulted in a more-complicated denominator. Specifically, the denominator of $Q(s)$ now contains the product of two polynomials of the generalized frequency $s$, suggesting that partial fractions will be useful to put this expression into a form for which the inverse Laplace transform is known.

One way to do that is to use the approach described in the tutorial on partial fractions on this book's website, which starts by writing each of the two polynomials as the denominator of a separate fraction:

$$\frac{V_s\omega_s/L}{(s^2 + \omega_s^2)[(s+a)^2 + \omega_1^2]} = \frac{V_s\omega_s}{L}\left[\frac{As+B}{s^2+\omega_s^2} + \frac{C(s+a)+D}{(s+a)^2+\omega_1^2}\right]. \quad (4.87)$$

You can solve this equation for the constants $A$, $B$, $C$, and $D$ in terms of the circuit parameters by multiplying both sides by the denominator of the left side and equating equal powers of $s$. That process leads to the following expressions for these constants:

$$A = \frac{-2a}{(\omega_s^2 - \omega_0^2)^2 + 4a^2\omega_s^2} = \frac{-R/L}{(\omega_s^2 - \frac{1}{LC})^2 + \left(\frac{R}{L}\right)^2\omega_s^2}, \quad (4.88)$$

$$B = \left(\frac{\omega_s^2 - \omega_0^2}{2a}\right)A, \quad (4.89)$$

$$C = -A, \quad (4.90)$$

and

$$D = -aA - B. \quad (4.91)$$

One of the chapter-end problems and its online solution provide details of the calculations leading to these results and contain expressions for $B$, $C$, and $D$ in terms of the circuit parameters.

With these constants in hand, you know the $s$-domain charge function $Q(s)$, and you can use the inverse Laplace transform to determine the time-domain function $q(t)$. To do that, start by taking the inverse Laplace transform of both sides of Eq. 4.86 and using Eq. 4.87:

$$q(t) = \mathcal{L}^{-1}[Q(s)] = \frac{V_s\omega_s}{L}\mathcal{L}^{-1}\left[\frac{As+B}{s^2+\omega_s^2} + \frac{C(s+a)+D}{(s+a)^2+\omega_1^2}\right]. \quad (4.92)$$

The linearity property of the Laplace transform means that you can apply the inverse transform to each term of $Q(s)$. Here is how that looks if you retain the notation $A$, $B$, $C$, and $D$ for the constants for simplicity:

$$\mathcal{L}^{-1}\left[\frac{As}{s^2+\omega_s^2}\right] = A\mathcal{L}^{-1}\left[\frac{s}{s^2+\omega_s^2}\right] = A\cos(\omega_s t), \quad (4.93)$$

$$\mathcal{L}^{-1}\left[\frac{B}{s^2+\omega_s^2}\right] = \left(\frac{B}{\omega_s}\right)\mathcal{L}^{-1}\left[\frac{\omega_s}{s^2+\omega_s^2}\right] = \left(\frac{B}{\omega_s}\right)\sin(\omega_s t), \quad (4.94)$$

$$\mathcal{L}^{-1}\left[\frac{C(s+a)}{(s+a)^2+\omega_1^2}\right] = C\mathcal{L}^{-1}\left[\frac{s+a}{(s+a)^2+\omega_1^2}\right] = Ce^{-at}\cos(\omega_1 t),$$

$$(4.95)$$

and

$$\mathcal{L}^{-1}\left[\frac{D}{(s+a)^2+\omega_1^2}\right] = \left(\frac{D}{\omega_1}\right)\mathcal{L}^{-1}\left[\frac{\omega_1}{(s+a)^2+\omega_1^2}\right]$$

$$= \left(\frac{D}{\omega_1}\right)e^{-at}\sin(\omega_1 t). \quad (4.96)$$

Adding these terms gives the time-domain charge function $q(t)$:

$$q(t) = \frac{V_s\omega_s}{L}\left[A\cos(\omega_s t) + \left(\frac{B}{\omega_s}\right)\sin(\omega_s t)\right.$$

$$\left. + Ce^{-at}\cos(\omega_1 t) + \left(\frac{D}{\omega_1}\right)e^{-at}\sin(\omega_1 t)\right]. \quad (4.97)$$

This equation tells you that the function $q(t)$ representing the capacitor charge in the driven series RLC circuit as a function of time consists of sinusoidal terms (both sine and cosine) at the drive frequency of $\omega_s$ as well as exponentially decreasing sinusoidal terms at the damped system frequency of $\omega_1$. The rate of decrease of those terms is determined by $a = R/2L$, so the larger the resistance (and the smaller the inductance), the faster the $\omega_1$ terms decrease.

Note also that the $\omega_s$ terms and the $\omega_1$ terms may be out of phase (that is, the peaks of one function may line up with the valleys of another), so these terms may partially cancel one another as they interfere destructively. Of course, if $\omega_1$ is significantly different from the drive frequency $\omega_s$, the phase relationship changes quickly with time, and the decreasing amplitude of the $\omega_1$ terms also changes the amount of cancellation even when the peaks temporarily align with valleys.

You can see this behavior in Figure 4.7, which shows the real and imaginary parts of $Q(s)$ as well as $q(t)$ for a series RLC circuit containing a 100-$\Omega$ resistor, 416.5-mH inductor, and 600-nF capacitor driven by an AC voltage source with $V_s = 10$ V at an angular frequency of $\omega_s = 2000$ rad/sec. With these parameters, the circuit is being driven at a frequency within about 3 rad/sec of the damped system frequency, and the attenuation frequency

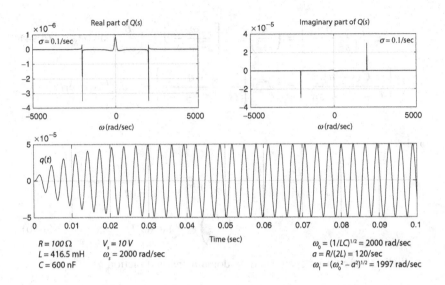

Figure 4.7  $Q(s)$ and $q(t)$ for series RLC circuit driven at $\omega_s = \omega_0$.

parameter is $a = 120/\text{sec}$, which means that the terms containing $e^{-at}$ are reduced to about 30% of their initial values after 0.01 seconds and 9% after 0.02 seconds.

This explains why the amplitude of the charge oscillations increases quickly after time $t = 0$ even though the presence of the resistor provides damping, which causes the amplitude to decrease in the undriven circuit described earlier in this section. The reason is that the out-of-phase terms in $q(t)$ containing the factor $e^{-at}$ decrease quickly over time, and the partial cancellation that occurs shortly after $t = 0$ is significantly reduced after 20 to 30 milliseconds.

The behavior of the circuit during the time that the exponentially decreasing terms contribute substantially to the total is called the "transient" behavior; at later times those terms become insignificant and the circuit then displays its "steady-state" behavior.

This analysis suggests that reducing the resistance $R$ while retaining the other parameters should slow the decrease in the $e^{-at}$ terms and extend the time over which the partial cancellation is effective in reducing the amplitude of the charge oscillation. In other words, the duration of the transient response should be increased.

Figure 4.8 shows the behavior of the circuit if the value of the resistor is reduced from 100 $\Omega$ to 10 $\Omega$, which reduces the value of $a$ from 120/sec to 12/sec. In this case, the $e^{-at}$ factor rolls off much more slowly; the amplitude

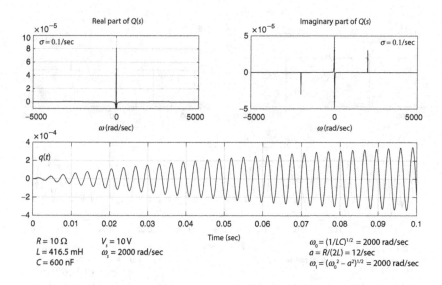

Figure 4.8  $Q(s)$ and $q(t)$ for series RLC circuit with smaller R driven at $\omega_0$.

of this factor is reduced only to about 89% of its initial value of unity after 0.01 seconds and 79% after 0.02 seconds. That means that the partial cancellation between the terms of $q(t)$ remains effective for a considerably longer time, as shown in the figure.

You may be wondering what happens if an RLC circuit is driven at a frequency significantly different from the natural or damped oscillation frequency. In that case, the phase relationship between the $\omega_s$ terms and the $\omega_1$ terms of $q(t)$ changes very quickly, and the combination of those terms can produce complicated behavior during the transient period. You can see one example of such behavior in Fig. 4.9, which is the response of the same series RLC circuit with $R = 100\,\Omega$, $L = 416.5$ mH, and $C = 600$ nF driven by a 10-V AC voltage source. But in this case that circuit is driven at a 25% higher frequency (2500 rad/sec rather than 2000 rad/sec). As you can see, the oscillation amplitude is greater than the steady-state amplitude for some portions of the transient period and smaller than the steady-state amplitude during others. This indicates that the interference between the terms of $q(t)$ is changing quickly from constructive to destructive, as expected when sinusoidal functions with significantly different frequencies are combined.

The functions $Q(s)$ and $q(t)$ just discussed pertain to the case in which $\omega_0^2 > a^2$, but the analysis is quite similar when that condition does not hold.

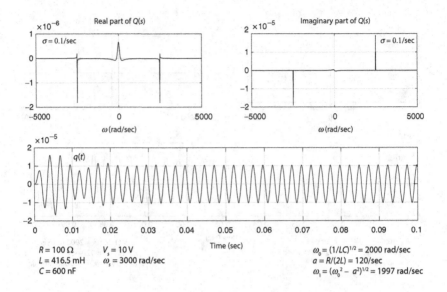

$R = 100\,\Omega$            $V_s = 10\,\text{V}$            $\omega_0 = (1/LC)^{1/2} = 2000\text{ rad/sec}$
$L = 416.5\text{ mH}$        $\omega_s = 3000\text{ rad/sec}$    $a = R/(2L) = 120/\text{sec}$
$C = 600\text{ nF}$                                $\omega_1 = (\omega_0^2 - a^2)^{1/2} = 1997\text{ rad/sec}$

Figure 4.9 $Q(s)$ and $q(t)$ for series RLC circuit driven above $\omega_0$.

If $\omega_0^2 < a^2$ you can find $Q(t)$ by reversing the terms in the definition of $\omega_1$, like this:

$$\omega_1^2 = \left(\frac{R}{2L}\right)^2 - \frac{1}{LC} = a^2 - \omega_0^2. \tag{4.98}$$

That gives

$$Q(s) = \frac{V_s\omega_s/L}{(s^2 + \omega_s^2)[(s + a)^2 - \omega_1^2]} \tag{4.99}$$

and

$$q(t) = \frac{V_s\omega_s}{L}\left[A\cos(\omega_s t) + \left(\frac{B}{\omega_s}\right)\sin(\omega_s t)\right.$$

$$\left. + Ce^{-at}\cosh(\omega_1 t) + \left(\frac{D}{\omega_1}\right)e^{-at}\sinh(\omega_1 t)\right]. \tag{4.100}$$

And if $\omega_0^2 = a^2$ then the $\omega_1^2 = \omega_0^2 - a^2$ term drops out of the denominator of $Q(s)$, leaving

$$Q(s) = \frac{V_s\omega_s/L}{(s^2 + \omega_s^2)(s + a)^2} \tag{4.101}$$

and $q(t)$ becomes

$$q(t) = \frac{V_s \omega_s}{L} \left[ A \cos (\omega_s t) + \left( \frac{B}{\omega_s} \right) \sin (\omega_s t) + C e^{-at} + D t e^{-at} \right]. \quad (4.102)$$

You can work through the analysis of an RLC circuit driven by a constant-voltage source in one of the chapter-end problems and its online solution.

## 4.4  Heat Flow

The application to which the Laplace transform is applied in this section is the flow of heat over space and time. Heat flow is the transfer of energy by thermal conduction or related processes, and as in the previous applications in this chapter, this process can be described by a differential equation. Hence the general approach will be to apply the Laplace transform to the differential equation in order to find the $s$-domain function $F(s)$ that solves the equation, and then to use the inverse Laplace transform to determine the time-domain solution $f(t)$.

There is, however, a very important difference between the heat equation and the equations describing the oscillations of a hanging mass or a series RLC circuit. As you may recall, the equations used in those examples are ordinary differential equations containing functions of a single variable (the position of the hanging mass $y(t)$ and the charge $q(t)$ in the RLC circuit). But the differential equation describing heat flow is not an ordinary differential equation, it is a partial differential equation that includes a first-order partial temporal derivative as well as a second-order partial spatial derivative.

To understand the origin and physical meaning of the differential equation pertaining to heat flow, often called the heat equation, a good place to begin is Fourier's law of thermal conduction. In the differential form of that law, the density of heat flow (that is, the heat flow per unit area) in any direction is proportional to the temperature change in that direction. Representing the density of heat flow as vector $\vec{q}$ with magnitude equal to the amount of heat energy passing through unit area in unit time (SI units of J/(m$^2$ sec) or W/m$^2$), Fourier's law looks like this:

$$\vec{q} = -\kappa \vec{\nabla} \tau, \quad (4.103)$$

in which the spatial change in temperature[3] is represented by the temperature gradient $\vec{\nabla} \tau$ (SI units of K/m), which is explained below. In this equation, $\kappa$

---

[3] The symbol $\tau$ is used for the time-domain function of temperature in this section, while $T$ is used for the corresponding $s$-domain function.

represents the thermal conductivity (SI units of J/(K m sec) or W/(K m), which is a measure of a material's ability to conduct heat – materials such as metals through which heat energy passes easily have high thermal conductivity, while insulators such as wood have low thermal conductivity.

The minus sign in Eq. 4.103 is important, because it makes this equation consistent with the observation that heat flows from regions of high temperature to regions of low temperature – that is, in the direction in which the change in temperature is negative.

Before showing how Fourier's law leads to the heat equation, it is worth a bit of effort to make sure you understand three uses of the differential operator represented by the symbol $\nabla$ (usually called "del" or "nabla"), all of which play a role in using the Laplace transform to solve the heat equation. The application of this differential operator can yield either a vector or a scalar result; for example, the gradient of a scalar function, written as $\vec{\nabla} f$, is a vector (having direction as well as magnitude), while the divergence of a vector function ($\vec{A}$), written as the dot product $\vec{\nabla} \circ \vec{A}$, is a scalar (having magnitude but no direction). Taking the divergence of a gradient ($\vec{\nabla} \circ \vec{\nabla}$) gives the Laplacian operator $\nabla^2$ (sometimes written as $\Delta$), which also produces a scalar result. In three-dimensional Cartesian coordinates, the gradient, divergence, and Laplacian are given by

$$\vec{\nabla} f(x, y, z) = \frac{\partial f}{\partial x} \hat{i} + \frac{\partial f}{\partial y} \hat{j} + \frac{\partial f}{\partial z} \hat{k} \qquad \text{gradient,}$$

$$\vec{\nabla} \circ \vec{A} = \frac{\partial A_x}{\partial x} + \frac{\partial A_y}{\partial y} + \frac{\partial A_z}{\partial z} \qquad \text{divergence,}$$

and

$$\nabla^2 f(x, y, z) = \frac{\partial^2 f}{\partial x^2} + \frac{\partial^2 f}{\partial y^2} + \frac{\partial^2 f}{\partial z^2} \qquad \text{Laplacian.}$$

So what is the physical meaning of each of these uses of the $\nabla$ operator? As suggested above, the vector gradient $\vec{\nabla}$ is a measure of how quickly and in what direction a function such as $f(x, y, z)$ changes over distance. So if $f(x, y, z)$ represents height above some reference level (such as mean sea level), the gradient $\vec{\nabla} f$ at any point on a hill tells you how steep the hill is at that location. Since it is a vector quantity, the gradient also has a direction, and that direction is the direction of steepest increase of the function. So in the case of ground height above a reference level, at every location the gradient vector points uphill, toward larger values of height.

To understand the physical meaning of the divergence $\vec{\nabla} \circ (\vec{A})$, consider a "vector field" such as a group of vectors specifying the speed and direction of the flow of water at various locations in a stream. The divergence of the vector field at any location is an indicator of whether the flow into a small volume at that location is greater than, equal to, or less than the flow out of that small volume. By convention, the divergence is taken to be positive if the flow out is greater than the flow in, and negative if the flow in is greater. This concept of "flow in vs. flow out" can be applied even when the vector field does not describe the actual flow of a physical material; for example, when specifying the magnitude and direction of an electric or magnetic field at various locations. But it is easy to visualize a divergence test in which you sprinkle a material such as sawdust into a stream and look for locations at which the particles "spread out" (positive divergence) or are pushed closer to one another (negative divergence).

Combining the divergence and the gradient to form the Laplacian ($\nabla^2$) finds use in many areas of physics and engineering, in part because the process of finding the locations of peaks and valleys of a function is useful in many applications. And how does the Laplacian, the divergence of the gradient of a function, help with that? One way to understand that is to consider the gradient vectors near the top of a circular hill. Since the gradient vectors all point uphill, if you look down on that hill from directly above, you'll see the gradient vectors "converging" on the hilltop. Hence the divergence of the gradient has a negative value at that location. Likewise, the gradient vectors all point uphill from the bottom of a circular depression or valley, so viewed from above those vectors will be seen diverging from the bottom, resulting in a positive value for divergence at that location.

Another useful perspective for this characteristic of the Laplacian concerns the value of a function at a location relative to its average value at nearby locations. Specifically, the Laplacian can be shown to be proportional to the difference between the value of a function at a specific location and the average value of the function at points surrounding that location at equal distance.[4] When the value of the function at a given location exceeds the average value of the function at surrounding points (as it does on at the top of a hill, where the height of surrounding points is less than the height at the peak), the Laplacian has a negative value. But at the bottom of a circular depression, the average height of the surrounding points is greater than the height at the bottom, so the Laplacian has a positive value at that point.

With some understanding of the gradient, divergence, and Laplacian, you should be ready to take the next step along the route that leads from Fourier's

---

[4] A detailed explanation of the relationship of the Laplacian to the average values of a function is provided in *A Student's Guide to Vectors and Tensors*.

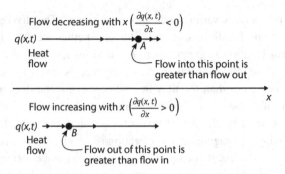

Figure 4.10 Effect of change in flow $q(x, t)$ over distance.

law to the heat equation. That step defines the variable $Q$ as the heat energy per unit volume (SI units of J/m$^3$), and considers the relationship of $Q$ to the flow of heat $\vec{q}$.

Consider the flow of heat shown in Fig. 4.10. In this figure, heat flow is one-dimensional along the $x$-axis, so $q(t)$ can be written as a scalar, taken as positive for heat flow to the right (toward positive $x$) and negative for flow to the left (toward negative $x$). The arrows in this diagram represent the heat-flow density at various locations, and the length of each arrow is proportional to the magnitude of the heat flow at that location. In the top portion of the figure, the arrows get shorter as you move to the right along the $x$-axis, meaning the heat flow is decreasing as $x$ increases. At a location such as the point marked "A" in the figure, the decreasing heat flow with $x$ means that the flow of heat into a small volume around point A will be greater than the flow out of that volume. In the absence of a source that adds energy or a sink that removes energy at point A, this means that the greater inward heat flow will cause the heat-energy density to increase over time at that location.

Now consider the lower portion of this figure, in which the heat flow is increasing in the positive-$x$ direction. In this case, the increasing heat flow with $x$ means that more energy will flow out of a small volume surrounding point B than flows in, and the greater outward heat flow will cause the heat-energy density at this location to decrease over time.

This relationship can be written mathematically using the partial time derivative of heat-energy density $Q$ and the divergence of the heat flow $\vec{q}$:

$$\frac{\partial Q}{\partial t} = -\vec{\nabla} \circ \vec{q},$$

in which the derivative of $Q$ is a partial derivative since $Q$ depends on both space $(x)$ and time $(t)$. Using the expression for $\vec{q}$ from Fourier's law (Eq. 4.103) makes this

$$\frac{\partial Q}{\partial t} = -\vec{\nabla} \circ (-\kappa \vec{\nabla} \tau) = \kappa \nabla^2 \tau. \tag{4.104}$$

The final step to the heat equation can be made by relating the change in heat energy of a material to the change in temperature of that material. That relationship is

$$\frac{\partial Q}{\partial t} = \rho c_p \frac{\partial \tau}{\partial t} \tag{4.105}$$

in which $\rho$ represents the density of the material (SI units of kg/m$^3$), $c_p$ represents the heat capacity of the material (SI units of J/(K kg)), and $\tau$ represents the temperature of the material in kelvins at the location of interest.

Inserting this expression into Eq. 4.104 gives

$$\rho c_p \frac{\partial \tau}{\partial t} = \kappa \nabla^2 \tau$$

or

$$\frac{\partial \tau}{\partial t} = \frac{\kappa}{\rho c_p} \nabla^2 \tau. \tag{4.106}$$

This is the heat equation, a partial differential equation which is a version of general equation called the "diffusion equation" that describes the flow or diffusion of a quantity over space and time. The diffusion equation has widespread application, with the diffusing quantity ranging from the concentration of a physical substance in fluid dynamics to probability in quantum mechanics, in which the Schrödinger equation is a version of the diffusion equation.

The law of diffusion can be written as

$$\frac{\partial f}{\partial t} = D \nabla^2 f, \tag{4.107}$$

in which $f$ represents the diffusing quantity and $D$ represents the diffusion coefficient. Following the logic described above for the $\nabla^2$ operator, you should be able to understand the physical meaning of the diffusion equation: in the absence of sources or sinks, the change in the function $f$ that describes a diffusing quantity over time at a specified location depends on the difference between the value of the function at that location and the average value of the

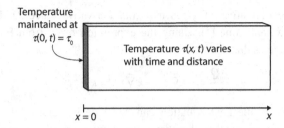

Figure 4.11  Material with temperature held at $\tau_0$ on face at $x = 0$.

function at surrounding points (that is, the Laplacian of the function). So if the value of the amount of a substance or the temperature of a material is greater than the average value at surrounding locations, diffusion causes the value to decrease over time, since the Laplacian is negative in that case. But if the value at a location is less than the value at surrounding points, diffusion causes the value to increase over time.

In one spatial dimension $(x)$, the diffusion equation is

$$\frac{\partial f(x,t)}{\partial t} = D\frac{\partial^2 f(x,t)}{\partial x^2} \qquad (4.108)$$

and, in this case, the diffusion equation tells you that the change over time of the function $f(x,t)$ is proportional to the curvature of that function over space, given by the second partial derivative with respect to $x$.

Getting back to the heat equation, here is how it looks in one spatial dimension:

$$\frac{\partial \tau(x,t)}{\partial t} = \frac{\kappa}{\rho c_p}\frac{\partial^2 \tau(x,t)}{\partial x^2}, \qquad (4.109)$$

and this is the equation to which the Laplace transform will be applied in the remainder of this section. The specific problem used as an example involves the flow of heat through a sample of material for which the left face is held at a constant temperature $\tau_0$, as illustrated in Fig. 4.11. As you can see, the material extends along the $x$-axis with the origin $(x = 0)$ at the left face. The material is taken as homogeneous (the same at all locations) and isotropic (the same in all directions within the material). The initial condition of constant temperature $\tau_0$ at the left face $(x = 0)$ can be expressed as $\tau(0,t) = \tau_0$.

Taking the Laplace transform of both sides of the one-dimensional heat equation (Eq. 4.109) results in

$$\mathcal{L}\left[\frac{\partial \tau(x,t)}{\partial t}\right] = \mathcal{L}\left[\frac{\kappa}{\rho c_p}\frac{\partial^2 \tau(x,t)}{\partial x^2}\right] = \frac{\kappa}{\rho c_p}\mathcal{L}\left[\frac{\partial^2 \tau(x,t)}{\partial x^2}\right]. \qquad (4.110)$$

The time derivative can be eliminated using the derivative property of the Laplace transform:

$$\frac{\partial \tau(x,t)}{\partial t} = sT(x,s) - \tau(x,0), \tag{4.111}$$

in which the $s$-domain temperature function $T(x,s)$ is the Laplace transform of the time-domain temperature function $\tau(x,t)$:

$$T(x,s) = \mathcal{L}[\tau(x,t)]. \tag{4.112}$$

The second-order partial spatial derivative of the time-domain function $\tau(x,t)$ can be converted to a second-order ordinary spatial derivative of the $s$-domain function $T(x,s)$, as explained in Section 4.1 (see Eq. 4.21):

$$\mathcal{L}\left[\frac{\partial^2 \tau(x,t)}{\partial x^2}\right] = \frac{d^2 T(x,s)}{dx^2}. \tag{4.113}$$

Inserting these expressions into Eq. 4.110 gives

$$sT(x,s) - \tau(x,0) = \frac{\kappa}{\rho c_p} \frac{d^2 T(x,s)}{dx^2} \tag{4.114}$$

or

$$\frac{d^2 T(x,s)}{dx^2} - \frac{\rho c_p}{\kappa}[sT(x,s) - \tau(x,0)] = 0. \tag{4.115}$$

So in this case the Laplace transform has not converted a differential equation into an algebraic equation, but it has produced a simpler ordinary differential equation that can be easily solved. If the initial temperature of the material is zero, so $\tau(x,0) = 0$, the solution to this equation is

$$T(x,s) = Ae^{\sqrt{\frac{\rho c_p}{\kappa}s}\,x} + Be^{-\sqrt{\frac{\rho c_p}{\kappa}s}\,x}, \tag{4.116}$$

as you can verify by substituting this expression for $T(x,s)$ into Eq. 4.115. But you know that Laplace transform functions must approach zero as $s$ approaches infinity, and the first term in this equation becomes infinite as $s$ approaches infinity. That means that the coefficient $A$ must equal zero, and

$$T(x,s) = Be^{-\sqrt{\frac{\rho c_p}{\kappa}s}\,x}. \tag{4.117}$$

To determine the constant $B$, insert $x = 0$ into this equation, giving

$$T(0,s) = Be^0 = B \tag{4.118}$$

and take the Laplace transform of the constant initial condition $\tau(0,t) = \tau_0$:

$$T(0,s) = \mathcal{L}[\tau(0,t)] = \mathcal{L}[\tau_0] = \tau_0 \mathcal{L}[1] = \tau_0 \frac{1}{s}. \tag{4.119}$$

Hence

$$T(0,s) = B = \tau_0 \frac{1}{s}, \tag{4.120}$$

which makes the expression for $T(x,s)$

$$T(x,s) = Be^{-\sqrt{\frac{\rho c_p}{\kappa} s} x} = \frac{\tau_0}{s} e^{-\sqrt{\frac{\rho c_p}{\kappa} s} x}. \tag{4.121}$$

This is the $s$-domain solution to the heat equation in this situation, and the inverse Laplacian transform of $T(x,s)$ gives the time-domain solution $\tau(x,t)$.

To find the inverse Laplace transform for this function, it helps to define a variable $a = x\sqrt{\rho c_p/\kappa}$, making $T(x,s)$ look like this:

$$T(x,s) = \frac{\tau_0}{s} e^{-\sqrt{\frac{\rho c_p}{\kappa} s} x} = \frac{\tau_0}{s} e^{-a\sqrt{s}}. \tag{4.122}$$

The inverse Laplace transform of a function of this form can be found in a table of Laplace transforms:

$$\mathcal{L}^{-1}\left[\frac{e^{-a\sqrt{s}}}{s}\right] = \text{erfc}\left(\frac{a}{2\sqrt{t}}\right), \tag{4.123}$$

in which "erfc" represents the complementary error function. This function is related to the error function "erf" obtained by integrating the Gaussian function $e^{-v^2}$:

$$\text{erf}(x) = \frac{2}{\sqrt{\pi}} \int_0^x e^{-v^2} dv. \tag{4.124}$$

Subtracting the error function from one gives the complementary error function:

$$\text{erfc}(x) = 1 - \text{erf}(x) = 1 - \left(\frac{2}{\sqrt{\pi}}\right) \int_0^x e^{v^2} dv. \tag{4.125}$$

The relationship of the Gaussian function to the error function and the complementary error function is illustrated in Fig. 4.12.

So the time-domain solution to the heat equation in this situation is

$$\tau(x,t) = \mathcal{L}^{-1}[T(x,s)] = \tau_0 \text{erfc}\left(\frac{x\sqrt{\rho c_p/\kappa}}{2\sqrt{t}}\right) = \tau_0 \text{erfc}\left(\frac{x\sqrt{\rho c_p}}{2\sqrt{\kappa t}}\right). \tag{4.126}$$

An example solution with constant temperature $\tau_0 = 300$ K at the left face and a material with the thermal properties of iron is shown in Fig. 4.13. In the upper portion of this figure, the real and imaginary parts of the

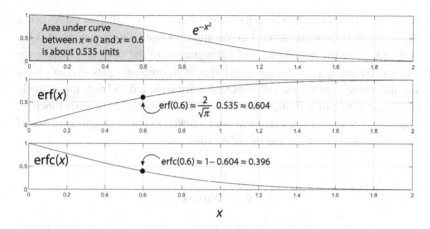

Figure 4.12 The Gaussian function $e^{-x^2}$, the error function (erf), and the complementary error function (erfc).

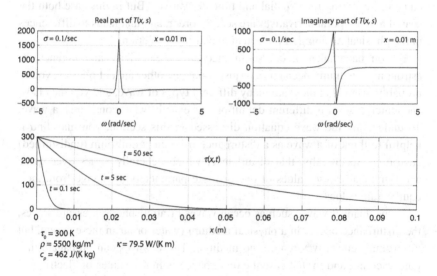

Figure 4.13 $T(x,s)$ at distance $x = 0.01$ m and $\tau(x,t)$ at times $t = 0.1$ sec, 5 sec, and 10 sec with the left face of the material held at $\tau_0$.

$s$-domain temperature function $T(x,s)$ are shown as functions of $\omega$ at distance $x = 0.01$ m, and in the lower portion the time-domain temperature $\tau(x,t)$ is shown as a function of $x$ at times $t = 0.1$ sec, 5 sec, and 50 sec.

As you may have anticipated, heat flow from the left end of the material, held at $\tau_0 = 300$ K in this example, causes the temperature at locations along the $x$-axis to rise over time from its initial value $\tau(x,0) = 0$. That makes sense physically, and the mathematical explanation for that rising temperature is the factor of $\sqrt{t}$ in the denominator of the argument of the complementary error function (since the erfc function gets larger as its argument gets smaller). Note also that the temperature at locations farther from the left end rises more slowly due to the factor of $x$ in the numerator of the complementary error function. You can work through this problem with a different initial condition in one of the chapter-end problems and its online solution.

## 4.5  Waves

In this section, you will see how the Laplace transform can be used to find solutions to the wave equation. Like the heat and diffusion equations discussed in the previous section, the wave equation is a second-order partial differential equation involving both spatial and time derivatives. But in this case both the spatial and temporal derivatives are second-order, and that leads to differences in the physical meaning of the equation and in its solutions.

If you have taken a course in mechanics, electricity and magnetism, astronomy, quantum mechanics, or just about any other area of physics, you're probably aware that there are many different types of wave, and you may have encountered several different definitions of exactly what constitutes a wave. To understand the wave equation discussed in this section, you may find it helpful to think of a wave as a disturbance from an equilibrium (undisturbed) condition. Specifically, that disturbance, which generally varies over space and time, causes the values of one or more parameters to change from their equilibrium values.

In mechanical waves such as ocean waves or atmospheric-pressure waves, the disturbance occurs in a physical medium (water or air in these cases), but electromagnetic waves require no medium. Those waves can propagate in a pure vacuum, and in that case the disturbance is in the values of electric and magnetic fields in the region in which the wave is propagating. You may see the disturbance of a wave called "displacement," but that doesn't necessarily mean that a physical material is being displaced, just that the wave is causing some parameter to be "displaced" in the sense that it moves away from its value in the equilibrium condition.

Note also that although many waves are periodic, meaning that the disturbance repeats itself over time (after one period) and over space (after one wavelength), other waves are not periodic. Those waves may consist of a disturbance such as a single pulse or a "wave packet" that contains constituent waves of different frequencies. Those component waves can interfere with one another constructively at some times and locations and destructively at others, producing a disturbance that is limited in time and space. But a true single-frequency wave such as a pure sine or cosine wave has uniform amplitude over all time, repeating itself over and over again, with no beginning and no end. Such single-frequency waves are a theoretical construct rather than a practical reality, but they form the basis of Fourier synthesis and analysis, as described in Chapter 1.

The amount of disturbance produced by a wave can be described as the "shape" of the wave, and you can see that shape by making a graph of the disturbance as a function of location or time. A mathematical function that represents the shape of a wave is often called a "wave function," sometimes written as the single word "wavefunction." A wave function can vary with both position and time, and the wave equation relates the spatial changes of the wave function to its temporal changes.

For reasons explained below, the most useful form of the wave equation for many applications involves second derivatives:

$$\frac{\partial^2 f}{\partial t^2} = v^2 \nabla^2 f, \tag{4.127}$$

in which $f$ represents the wave function, $v$ is the speed of propagation of the wave, and $\nabla^2$ is the Laplacian operator described in the previous section.

Expressing the Laplacian in three-dimensional Cartesian coordinates makes the wave equation

$$\frac{\partial^2 f(x,y,z,t)}{\partial t^2} = v^2 \left[ \frac{\partial^2 f(x,y,z,t)}{\partial x^2} + \frac{\partial^2 f(x,y,z,t)}{\partial y^2} + \frac{\partial^2 f(x,y,z,t)}{\partial z^2} \right]$$
$$\tag{4.128}$$

and in the one-dimensional case it looks like this:

$$\frac{\partial^2 f(x,t)}{\partial t^2} = v^2 \frac{\partial^2 f(x,t)}{\partial x^2}. \tag{4.129}$$

To understand the origin and meaning of this form of the wave equation, recall that the first temporal derivative $\partial f(x,t)/\partial t$ represents the instantaneous slope of a function over time, and the second temporal derivative $\partial^2 f(x,t)/\partial t^2$ tells you how quickly the slope of the function $f(x,t)$ changes with time. When you make a graph of a function that has a slope that changes over time, the line representing the function on that graph is curved, and the second temporal derivative is a measure of the curvature of the function over time. Likewise, the second partial derivative $\partial^2 f(x,t)/\partial x^2$ tells you the change in the slope of the function $f(x,t)$ over distance ($x$), and that change in slope is a measure of the curvature of the function with distance. Hence the wave equation says that the wave function's curvature in time is proportional to its curvature over space, and the constant of proportionality is the speed ($v$) of the wave squared.

The general form of a wave function satisfying the wave equation for a wave propagating in the positive-$x$ direction is $f(kx - \omega t)$, in which the function $f$ describes the shape of the wave. In case you need a bit of a refresher on wave parameters, in this expression the variable "$k$" represents the wavenumber, which is inversely proportional to the wavelength of the wave. Specifically, for a wave with wavelength $\lambda$ (SI units of meters) and angular frequency $\omega$ (SI units of radians per second), the wavenumber $k$ is defined as

$$k = \frac{2\pi}{\lambda} \tag{4.130}$$

and has SI units of radians per meter. Similarly, the angular frequency $\omega$ is defined as

$$\omega = \frac{2\pi}{T}, \tag{4.131}$$

in which $T$ represents the period of the wave (SI units of seconds). In the wave function with the form of $f(kx - \omega t)$, the role of $k$ in the spatial domain is similar to the role of $\omega$ in the time domain. That is because multiplying the time $t$ by the angular frequency $\omega$ tells you how many radians of phase the wave goes through in time $t$, and multiplying the distance $x$ by the wavenumber $k$ tells you how many radians of phase the wave goes through in distance $x$. Hence $k$ converts distance to phase in the same way that $\omega$ converts time to phase.

One additional point about $\omega$ and $k$: the ratio of these parameters ($\omega/k$) has dimensions of speed (SI units of meters per second); the "phase speed" given by this ratio is the speed $v$ at which a given phase point (for example, a peak, zero crossing, or valley of a sinusoidal wave) propagates. For nondispersive waves, that is, waves in which all frequency components propagate with the same speed, this is also the speed with which the envelope of the wave moves.

Although it is possible to write a version of the wave equation using first-order rather than second-order partial derivatives, the second-order wave equation given by Eq. 4.129 has the advantage of applying to waves propagating in both the positive-$x$ and the negative-$x$ direction.[5]

And how does the Laplace transform help you solve the second-order partial differential wave equation? To understand that, consider the situation shown in Fig. 4.14. As depicted in this figure, a string extending from $x = 0$ to

---

[5] The reason for this is explained in detail in A Student's Guide to Waves.

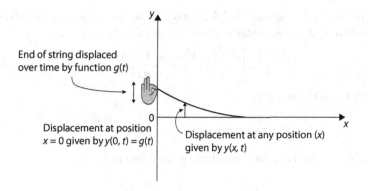

Figure 4.14 String driven by function $g(t)$.

$x = +\infty$ is disturbed from its equilibrium position by a force (such as the hand shown in the figure). The applied force causes the left end of the string to undergo a physical displacement over time, and that displacement is given by the time-domain function $g(t)$. Calling the vertical displacement of the string a function of space and time $y(x,t)$ makes the one-dimensional wave equation look like this:

$$\frac{\partial^2 y(x,t)}{\partial t^2} = v^2 \frac{\partial^2 y(x,t)}{\partial x^2}. \tag{4.132}$$

As you may have anticipated if you've worked through the previous applications in this chapter, the boundary conditions have a significant impact on the solution of differential equations such as the wave equation. Taking the initial time conditions as zero displacement and zero vertical speed at time $t = 0$ and the spatial boundary condition that the displacement of the left end of the string is given by the function $g(t)$, the boundary conditions are

$$y(x,0) = 0 \qquad \frac{dy}{dt}\bigg|_{t=0} = 0 \qquad y(0,t) = g(t). \tag{4.133}$$

With these boundary conditions in hand, along with condition that the displacement of the string must remain finite for all time over all locations, the next step is to take the Laplace transform of both sides of Eq. 4.132:

$$\mathcal{L}\left[\frac{\partial^2 y(x,t)}{\partial t^2}\right] = \mathcal{L}\left[v^2 \frac{\partial^2 y(x,t)}{\partial x^2}\right] = v^2 \mathcal{L}\left[\frac{\partial^2 y(x,t)}{\partial x^2}\right]. \tag{4.134}$$

The time-derivative property of the Laplace transform tells you that

$$\mathcal{L}\left[\frac{\partial^2 y(x,t)}{\partial t^2}\right] = s^2 Y(x,s) - sy(x,0) - \frac{\partial y}{\partial t}\bigg|_{t=0} \tag{4.135}$$

and the discussion around Eqs. 4.20 and 4.21 in Section 4.1 explains how to relate first- and second-order $x$-derivatives in the $t$- and $s$-domains:

$$\mathcal{L}\left[\frac{\partial^2 y(x,t)}{\partial x^2}\right] = \frac{d^2 Y(x,s)}{dx^2}. \tag{4.136}$$

Hence Eq. 4.134 becomes

$$s^2 Y(x,s) - s y(x,0) - \frac{\partial y}{\partial t}\Big|_{t=0} = v^2 \frac{d^2 Y(x,s)}{dx^2} \tag{4.137}$$

and inserting the boundary conditions given above makes this

$$s^2 Y(x,s) - 0 - 0 = v^2 \frac{d^2 Y(x,s)}{dx^2} \tag{4.138}$$

or

$$\frac{d^2 Y(x,s)}{dx^2} - \frac{s^2}{v^2} Y(x,s) = 0. \tag{4.139}$$

At first glance, this equation looks almost identical to the the differential equation for heat flow (Eq. 4.115) discussed in the previous section. But a closer inspection reveals an important difference: in this equation, the factor in front of the $s$-domain function $Y(x,s)$ is $s^2$ rather than $s$. That difference means that the general solution includes exponential terms containing $s$ rather than $\sqrt{s}$:

$$Y(x,s) = A e^{\frac{s}{v}x} + B e^{-\frac{s}{v}x} \tag{4.140}$$

and the requirement that $Y(x,s)$ remain bounded as $s$ goes to infinity means that the coefficient $A$ must be zero. So

$$Y(x,s) = B e^{-\frac{s}{v}x}. \tag{4.141}$$

As in the previous example, the coefficient $B$ can be determined by inserting $x = 0$ into this equation, which gives

$$Y(0,s) = B e^0 = B. \tag{4.142}$$

Note also that the $s$-domain function $Y(0,s)$ is the Laplace transform of the time-domain function $y(0,t)$:

$$Y(0,s) = \mathcal{L}[y(0,t)], \tag{4.143}$$

and that the displacement $y(0,t)$ at the left end of the string ($x = 0$) is specified above to be the driving function $g(t)$. Denoting the Laplace transform of $g(t)$ as $G(s)$, this means that

$$Y(0,s) = \mathcal{L}[y(0,t)] = \mathcal{L}[g(t)] = G(s), \tag{4.144}$$

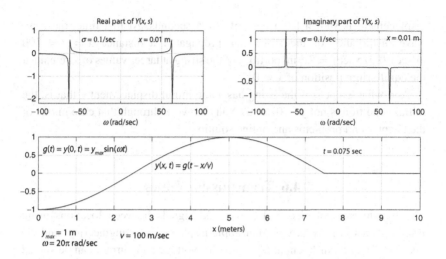

Figure 4.15 $Y(x,s)$ and $y(x,t)$ for string wave driven by $g(t) = y_{max} \sin(\omega t)$.

and Eq. 4.142 says that $Y(0,s) = B$. Setting $B$ equal to $G(s)$ makes Eq. 4.141

$$Y(x,s) = Be^{-\frac{s}{v}x} = G(s)e^{-\frac{s}{v}x}. \qquad (4.145)$$

So this is the $s$-domain function that satisfies the wave equation and the boundary conditions, and as in the previous examples, the final step is to get back to the time domain by taking the inverse Laplace transform of $Y(x,s)$. You know that $g(t)$ is the inverse Laplace transform of $G(s)$, and the time-shift property of the Laplace transform tells you that multiplying by an exponential in the $s$-domain results in a time shift in the $t$-domain. Thus

$$y(x,t) = \mathcal{L}^{-1}[Y(x,s)] = \mathcal{L}^{-1}[G(s)e^{-\frac{s}{v}x}] = g\left(t - \frac{x}{v}\right) \qquad (4.146)$$

for all time $t > x/v$.

This is the solution for the displacement as a function of $x$ and $t$, and it tells you that at any location $x$, the string undergoes displacement that equals the displacement at the string's left end, given by $g(t)$ and delayed by the time $x/v$ that it takes for the disturbance, traveling at the wave speed $v$, to propagate from $x = 0$ to that point.

An example of this solution is shown in Fig. 4.15. In this example, the function driving the left end of the string is a sine function with angular frequency of $20\pi$ rad/sec and maximum vertical displacement of 1 m. The wave speed in the string is taken to be 100 m/sec, and the plot of the string displacement is shown for a time of 0.075 seconds. At that time, the sine-wave disturbance of the string has propagated a distance $x = vt = (100$ m/sec)(0.075 sec) $= 7.5$ m; portions of the string at larger values of $x$ remain at the equilibrium position of $y = 0$.

And what happens if the string has some initial displacement – that is, the initial condition is not $y(x, 0) = 0$? You can work through that case in one of the chapter-end problems and online solutions.

## 4.6 Transmission Lines

The transmission of electrical power and signals between locations using transmissions lines is an important topic in physics and engineering, and the resulting coupled differential equations for voltage and current can be solved using the Laplace transform. There are many different kinds of transmission lines such as coaxial cable, twisted-pair, and waveguide, but a common characteristic of all types is that the properties of the transmission line are distributed along the line, rather than in the "lumped elements" that are the traditional devices such as resistors, capacitors, and inductors that make up an electrical circuit.

The properties of a transmission line include resistance, capacitance, inductance, and conductance, usually represented by the symbols $R, C, L$, and $G$, with one very important difference: in this case, the values of these parameters are typically specified as the amount per unit length rather than the total quantity. So when you're analyzing a transmission line, the SI units of $R$ are not simply ohms, but ohms per meter. Likewise, $C$ represents the capacitance per unit length (farads per meter in the SI system), $L$ represents the inductance per unit length (SI units of henries per meter), and $G$ represents the conductance per unit length (SI units of inverse ohms or siemens (sometimes called "mhos") per meter). The authors of some texts use slightly modified symbols to represent these "per unit length" quantities, such as $R', C', L'$, and $G'$, but most just use the same letters and expect you to remember the units. Why is this worth mentioning? One reason is that an important way to check your work and the expressions you derive when working a problem is to make sure that your equations and answers have the proper dimensions, and that means that you must include the inverse meters in the units of $R, C, L$, and $G$ when working transmission-line problems.

So how do these properties of a transmission line affect the response of the line to an applied voltage? To understand that, take a look at the sketch of a transmission line in Fig. 4.16.

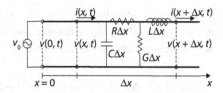

Figure 4.16 Voltage ($v$) and current ($i$) for transmission line with distributed resistance ($R$), inductance ($L$), capacitance ($C$), and conductance ($G$).

This figure depicts a semi-infinite transmission line extending from $x = 0$ to $x = \infty$. The two dark horizontal lines represent the conductors across which a voltage is applied at the "sending" end (the left end in this figure); the voltage between these two conductors at any location and time is given by $v(x,t)$ and the current is given by $i(x,t)$. In this section, lowercase letters $v(x,t)$ and $i(x,t)$ are used for time-domain voltage and current, while uppercase letters $V(x,s)$ and $I(x,s)$ are used for generalized frequency-domain voltage and current. So

$$V(x,s) = \mathcal{L}[v(x,t)] \qquad I(x,s) = \mathcal{L}[i(x,t)].$$

As shown in the figure, the applied voltage at the sending end is $v_0 = v(0,t)$, and over the distance $\Delta x$ the voltage changes from $v(x,t)$ to $v(x + \Delta x,t)$ while the current changes from $i(x,t)$ to $i(x + \Delta x,t)$. The amount by which the voltage and current change over space and time is determined by the line's resistance (which impedes the flow of current along the conductors), the capacitance between the two conductors (which causes energy to be stored in the electric field between the conductors), the inductance (which causes energy to be stored in the magnetic field around the conductors), and the conductance (which allows leakage current to flow between the conductors).

The voltage change along the line depends on the line's resistance and inductance as well as the current and the change in current with time. If the voltage change between locations $x$ and $x + \Delta x$ is defined as

$$\Delta_x v(x,t) = v(x + \Delta x,t) - v(x,t) = \frac{\partial v(x,t)}{\partial x} \Delta x, \qquad (4.147)$$

then

$$\Delta_x v(x,t) = \frac{\partial v(x,t)}{\partial x} \Delta x = -\frac{\partial i(x,t)}{\partial t} L \Delta x - i(x,t) R \Delta x, \qquad (4.148)$$

in which $L\Delta x$ is the inductance and $R\Delta x$ is the resistance of a portion of the line of length $\Delta x$.

Likewise, the current change along the line depends on the line's capacitance and conductance as well as the voltage and the change in voltage with time. Defining the current change between locations $x$ and $x + \Delta x$ as

$$\Delta_x i(x,t) = i(x + \Delta x, t) - i(x,t) = \frac{\partial i(x,t)}{\partial x} \Delta x \qquad (4.149)$$

then

$$\Delta_x i(x,t) = \frac{\partial i(x,t)}{\partial x} \Delta x = -\frac{\partial v(x,t)}{\partial t} C \Delta x - v(x,t) G \Delta x, \qquad (4.150)$$

in which $C \Delta x$ is the capacitance and $G \Delta x$ is the conductance of a portion of the line of length $\Delta x$.

Now consider a transmission line in which the inductance ($L$) and conductance ($G$) per meter are negligible, with boundary conditions of zero initial voltage and current at time $t = 0$, so $v(x,0) = 0$ and $i(x,0) = 0$. As shown in the figure, the voltage applied by the generator at the sending end of the line ($x = 0$) is $v(0,t) = v_0$. One final boundary condition is that $v(x,t)$ must remain bounded at all locations and times.

Under these conditions, Eq. 4.148 becomes

$$\frac{\partial v(x,t)}{\partial x} = -i(x,t)R \qquad (4.151)$$

and Eq. 4.150 becomes

$$\frac{\partial i(x,t)}{\partial x} = -\frac{\partial v(x,t)}{\partial t} C. \qquad (4.152)$$

As you have probably anticipated if you've worked through the other sections of this chapter, the next step is to take the Laplace transform of both of these equations. That looks like this:

$$\mathcal{L}\left[\frac{\partial v(x,t)}{\partial x}\right] = \mathcal{L}[-i(x,t)R] = -R\mathcal{L}[i(x,t)] \qquad (4.153)$$

and

$$\mathcal{L}\left[\frac{\partial i(x,t)}{\partial x}\right] = \mathcal{L}\left[-\frac{\partial v(x,t)}{\partial t}C\right] = -C\mathcal{L}\left[\frac{\partial v(x,t)}{\partial t}\right]. \qquad (4.154)$$

As mentioned above, the Laplace transform of the time-domain voltage function is $\mathcal{L}[v(x,t)] = V(x,s)$ and the Laplace transform of the time-domain current function is $\mathcal{L}[i(x,t)] = I(x,s)$. Applying the $x$-derivative property of the Laplace transform to Eq. 4.153 gives

$$\mathcal{L}\left[\frac{\partial v(x,t)}{\partial x}\right] = \frac{dV(x,s)}{dx} = -R\mathcal{L}[i(x,t)] = -RI(x,s)$$

or

$$\frac{dV(x,s)}{dx} = -RI(x,s) \tag{4.155}$$

and applying the both the $x$-derivative and the $t$-derivative properties of the Laplace transform to Eq. 4.154 gives

$$\mathcal{L}\left[\frac{\partial i(x,t)}{\partial x}\right] = \frac{dI(x,s)}{dx} = -C\mathcal{L}\left[\frac{\partial v(x,t)}{\partial t}\right] = -C\left[sV(x,s) - v(x,0)\right].$$

$$\tag{4.156}$$

Applying the initial condition $v(x,0) = 0$ simplifies the last equation to

$$\frac{dI(x,s)}{dx} = -C\left[sV(x,s)\right]. \tag{4.157}$$

Now take the derivative with respect to $x$ of both sides of Eq. 4.155:

$$\frac{d}{dx}\left[\frac{dV(x,s)}{dx}\right] = \frac{d}{dx}[-RI(x,s)]$$

$$\frac{d^2V(x,s)}{dx^2} = -R\frac{dI(x,s)}{dx} \tag{4.158}$$

and substitute the expression for the $x$-derivative of $I(x,s)$ from Eq. 4.157. That gives

$$\frac{d^2V(x,s)}{dx^2} = -R[-CsV(x,s)] = RCsV(x,s). \tag{4.159}$$

Just as in the heat-flow application discussed in Section 4.4, the solution to this second-order ordinary differential equation can be written as

$$V(x,s) = Ae^{\sqrt{RCs}x} + Be^{-\sqrt{RCs}x}. \tag{4.160}$$

But once again, the Laplace transform $V(x,s)$ must remain bounded for all $x$, so the coefficient $A$ of the first term must equal zero, and

$$V(x,s) = Be^{-\sqrt{RCs}x}. \tag{4.161}$$

As in the heat-flow application, the remaining coefficient can be determined using the boundary condition on $v(0,t) = v_0$ which means that $\mathcal{L}[v(0,t)] = \mathcal{L}[v_0]$. And since $v_0$ is a constant, you know that $\mathcal{L}[v_0] = \frac{v_0}{s}$, so

$$V(0,s) = \frac{v_0}{s} = Be^0 = B, \tag{4.162}$$

which makes Eq. 4.161 look like this:

$$V(x,s) = Be^{-\sqrt{RCs}x} = \frac{v_0}{s}e^{-\sqrt{RCs}x}. \tag{4.163}$$

As shown in Section 4.4, the inverse Laplace transform of a function of this form is

$$\mathcal{L}^{-1}\left[\frac{e^{-a\sqrt{s}}}{s}\right] = \text{erfc}\frac{a}{2\sqrt{t}} \tag{4.123}$$

and letting $a = \sqrt{RC}x$ gives

$$v(x,t) = \mathcal{L}^{-1}[V(x,s)] = v_0\mathcal{L}^{-1}\left[\frac{e^{-\sqrt{RCs}x}}{s}\right] = (v_0)\text{erfc}\left(\frac{\sqrt{RC}x}{2\sqrt{t}}\right), \tag{4.164}$$

which is the desired solution to the differential equation for the voltage as a function of location and time for the given boundary conditions.

To determine the current $i(x,t)$ along the transmission line, note that solving Eq. 4.151 for the current gives

$$i(x,t) = -\frac{1}{R}\frac{\partial v(x,t)}{\partial x}. \tag{4.165}$$

To take the partial derivative of $v(x,t)$ given by Eq. 4.164, recall that the complementary error function is related to the error function by

$$\text{erfc}\left(\frac{\sqrt{RC}x}{2\sqrt{t}}\right) = 1 - \text{erf}\left(\frac{\sqrt{RC}x}{2\sqrt{t}}\right) \tag{4.166}$$

and the error function in this case is

$$\text{erf}\left(\frac{\sqrt{RC}x}{2\sqrt{t}}\right) = \frac{2}{\sqrt{\pi}}\int_0^{\sqrt{RC}x/2\sqrt{t}} e^{-v^2}dv. \tag{4.167}$$

So taking the partial derivative of $v(x,t)$ with respect to $x$ looks like this:

$$\frac{\partial v(x,t)}{\partial x} = v_0 \frac{\partial}{\partial x}\left[1 - \text{erf}\left(\frac{\sqrt{RC}x}{2\sqrt{t}}\right)\right]$$

$$= v_0 \frac{\partial}{\partial x}\left[1 - \frac{2}{\sqrt{\pi}}\int_0^{\sqrt{RC}x/2\sqrt{t}} e^{-v^2}dv\right], \qquad (4.168)$$

which may appear a bit daunting at first look. But recall from the fundamental theorem of calculus that

$$\frac{\partial}{\partial x}\left[\int_0^{a(x)} f(v)dv\right] = f(a)\frac{\partial a(x)}{\partial x}, \qquad (4.169)$$

which you can apply to this case by setting $a(x) = \sqrt{RC}x/(2\sqrt{t})$ and $f(v) = e^{-v^2}$. Doing that gives

$$\frac{\partial v(x,t)}{\partial x} = v_0\frac{\partial(1)}{\partial x} - v_0\frac{2}{\sqrt{\pi}}e^{-\left(\frac{\sqrt{RC}x}{2\sqrt{t}}\right)^2}\frac{\partial}{\partial x}\left(\frac{\sqrt{RC}x}{2\sqrt{t}}\right)$$

$$= 0 - \frac{2v_0}{\sqrt{\pi}}e^{-\frac{RCx^2}{4t}}\left(\frac{\sqrt{RC}}{2\sqrt{t}}\right) = -v_0\sqrt{\frac{RC}{\pi t}}e^{-\frac{RCx^2}{4t}}. \qquad (4.170)$$

Thus the current $i(x,t)$ is given by

$$i(x,t) = -\frac{1}{R}\frac{\partial v(x,t)}{\partial x} = -\frac{1}{R}\left(-v_0\sqrt{\frac{RC}{\pi t}}e^{-\frac{RCx^2}{4t}}\right)$$

$$= v_0\sqrt{\frac{C}{R\pi t}}e^{-\frac{RCx^2}{4t}}. \qquad (4.171)$$

As you may have anticipated from the similarity of the solution for the voltage $v(x,t)$ to the solution for the temperature $\tau(x,t)$ in Section 4.4, the voltage $v_0$ applied at time $t = 0$ to the sending end of the transmission line causes the voltage along the line to increase from its initial value over time, asymptotically approaching $v_0$ as $t \to \infty$.

In the top portion of Fig. 4.17, you can see the real and imaginary parts of the s-domain function $V(x,s)$ for a transmission line with the resistance and capacitance per unit length of the coaxial cable type RG-58/U transmission line ($R = 0.053$ $\Omega$/m and $C = 93.5$ pF/m) with applied voltage $v_0 = 10$ V. The middle and bottom portions of this figure show the time-domain voltage $v(x,t)$ and current $i(x,t)$ for times $t = 1$ nsec, 10 nsec, and 100 nsec after the voltage is applied.

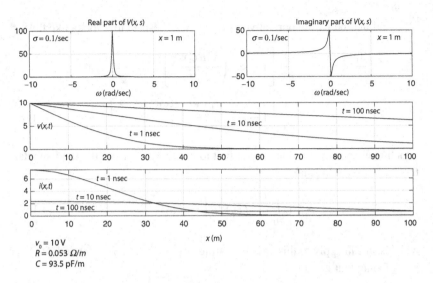

Figure 4.17 Example solutions $V(x,s)$ and $v(x,t)$ for the RG-58/U transmission line driven by constant voltage $v_0 = 10$ V.

As always, I strongly recommend that you work through the problems in the final section of this chapter – and don't forget that help is available in the form of full, interactive solutions on this book's website.

## 4.7 Problems

1. Use partial fractions to decompose the $s$-domain function $F(s)$ in Eq. 4.10 into the five terms shown in Eq. 4.11.

2. Use the Laplace transform and the inverse Laplace transform to solve the differential equation

$$\frac{d^2 f(t)}{dt^2} + 4\frac{df(t)}{dt} - 2f(t) = 5e^{-t}$$

with initial conditions $f(0) = -3$ and $\frac{df(t)}{dt}|_{t=0} = 4$.

3. Here is a differential equation with nonconstant coefficients:

$$t\frac{d^2 f(t)}{dt^2} + t\frac{df(t)}{dt} + f(t) = 0.$$

Find $f(t)$ using the Laplace transform and the inverse Laplace transform if the initial conditions are $f(0) = 0$ and $\frac{df(t)}{dt}|_{t=0} = 2$.

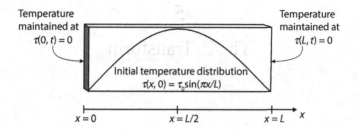

Temperature maintained at $\tau(0, t) = 0$

Temperature maintained at $\tau(L, t) = 0$

Initial temperature distribution $\tau(x, 0) = \tau_o \sin(\pi x/L)$

$x = 0$  $x = L/2$  $x = L$  $x$

4. Use the Laplace transform to convert the following partial differential equations into ordinary differential equations and solve those equations:

(a) $\frac{\partial f(x,t)}{\partial t} = \frac{\partial f(x,t)}{\partial x} + f(x,t)$ with $f(x,0) = 4e^{-2x}$.

(b) $\frac{\partial f(x,t)}{\partial t} = \frac{\partial f(x,t)}{\partial x}$ with $f(x,0) = \cos(bx)$.

5. Show that the expression for $y(t)$ in Eq. 4.54 is a solution to the differential equation of motion for a mass hanging on a spring (Eq. 4.35) and satisfies the initial conditions $y(0) = y_d$ and $dy/dt = 0$ at time $t = 0$.

6. Derive the expressions shown in Eqs. 4.88 to 4.91 for the partial-fraction decomposition of Eq. 4.87 and write the constants $B, C,$ and $D$ in terms of circuit parameters.

7. Find the $s$-domain function $F(s)$ and the time-domain function $f(t)$ for a series RLC circuit in which the voltage source is a battery with constant emf $V_0$ and zero initial charge and current.

8. Find the $s$-domain temperature function $T(x,s)$ and the time-domain temperature function $\tau(x,t)$ for the block of material shown above if the ends at $x = 0$ and $x = L$ are held at temperature $\tau = 0$ and the initial temperature distribution is $\tau(x,0) = \tau_0 \sin\left(\frac{\pi x}{L}\right)$.

9. Find the $s$-domain function $Y(s)$ and the time-domain function $y(t)$ for the string wave discussed in Section 4.5 if the initial displacement of the string at time $(t = 0)$ is $y(x,0) = y_0 \sin(ax)$.

10. Find the time-domain voltage $v(x,t)$ and current $i(x,t)$ for the transmission line discussed in Section 4.6 if the line has an initial voltage $v(x,0) = v_i$ in which $v_i$ is a constant.

# 5

# The Z-Transform

Another useful transform related to the Fourier and Laplace transforms is the Z-transform, which, like the Laplace transform, converts a time-domain function into a frequency-domain function of a generalized complex frequency parameter. But the Z-transform operates on sampled (or "discrete-time") functions, often called "sequences" while the Laplace transform operates on continuous-time functions. Thus the relationship between the Z-transform and the Laplace transform parallels the relationship between the discrete-time Fourier transform and the continuous-time Fourier transform. Understanding the concepts and mathematics of discrete-time transforms such as the Z-transform is especially important for solving problems and designing devices and systems using digital computers, in which differential equations become difference equations and signals are represented by sequences of data values.

You can find comprehensive discussions of the Z-transform in several textbooks and short introductions on many websites, so rather than duplicating that material, this chapter is designed to provide a bridge to the Z-transform from the discussion and examples of the Laplace transform and its characteristics in previous chapters.

The foundations of that bridge are laid in Section 5.1 with the definition of the Z-transform and an explanation of its relationship to the Laplace transform, along with graphical depictions of the mapping of the Laplace $s$-plane to the $z$-plane and a description of the inverse Z-transform. Section 5.2 has examples of the Z-transform of several basic discrete-time functions, and the properties of the Z-transform are presented in Section 5.3. As in previous chapters, the final section of this chapter contains a set of problems which will allow you to check your understanding of the chapter's material.

## 5.1 Introduction to the Z-Transform

A sampled time-domain function can be written as the product of a continuous-time function such as $f(t)$ and series of delta ($\delta$) functions:

$$f_{samp}(t) = \sum_{n=-\infty}^{\infty} f(t)\delta(t - nT_0), \qquad (5.1)$$

in which $T_0$ is the sample period and $f_{samp}(t)$ is a sampled version of the continuous time-domain function $f(t)$. Since $\delta(t - nT_0)$ has nonzero value only at times $t = nT_0$, this can be written as

$$f_{samp}(t) = \sum_{n=-\infty}^{\infty} f(nT_0)\delta(t - nT_0) \qquad (5.2)$$

and $f_{samp}(t)$ can be represented by a sequence of values. Note that $n$ takes on negative as well as positive values in this case, but in many applications only positive-time samples are used. In that case, the sampled function can be defined as

$$f_{samp}(t) = \sum_{n=-\infty}^{\infty} f(t)u(t)\delta(t - nT_0) = \sum_{n=0}^{\infty} f(nT_0)\delta(t - nT_0), \qquad (5.3)$$

in which $u(t)$ represents the unit-step function with a value of zero for $t < 0$ and unity for $t \geq 0$. These sampled positive-time functions will be the primary focus of the Z-transform discussions in this chapter.

### Relationship to the Laplace Transform

To understand the origin of the Z-transform, start by considering the Laplace transform of the sampled time-domain function of Eq. 5.3. That transform can be written as

$$\mathcal{L}[f_{samp}(t)] = \mathcal{L}\left[\sum_{n=0}^{\infty} f(nT_0)\delta(t - nT_0)\right]$$

$$= \int_{0^-}^{\infty} \sum_{n=0}^{\infty} f(nT_0)\delta(t - nT_0)e^{-st}dt. \qquad (5.4)$$

But each value of $f(nT_0)$ is a constant, so these terms in the summation can be moved outside the integral over time:

$$\mathcal{L}[f_{samp}(t)] = \sum_{n=0}^{\infty} f(nT_0) \int_{0^-}^{\infty} \delta(t - nT_0)e^{-st} dt$$

$$= \sum_{n=0}^{\infty} f(nT_0)e^{-snT_0}, \tag{5.5}$$

in which the sifting property of the time-shifted Dirac $\delta$-function described in Chapter 1 has been used.

Defining a new generalized-frequency variable

$$z = e^{sT_0} = e^{(\sigma+i\omega)T_0} \tag{5.6}$$

makes Eq. 5.5

$$\mathcal{L}[f_{samp}(t)] = \sum_{n=0}^{\infty} f(nT_0)z^{-n} = F(z), \tag{5.7}$$

which is the definition of the one-sided Z-transform of the discrete-time function $f(nT_0)$, with $F(z)$ representing the $z$-domain function that results from the process.

In many texts, the sample period $T_0$ is set to one unit of whatever time units are convenient for the situation. Setting the sample period $T_0$ to unity makes the unilateral Z-transform look like this:

$$F(z) = \mathcal{L}[f_{samp}(t)] = \sum_{n=0}^{\infty} f(n)z^{-n}, \tag{5.8}$$

and just as the symbol $\mathcal{L}$ is used to denote the process of taking the Laplace transform of a time-domain function, the Z-transformation operator can be represented by the symbol $\mathcal{Z}$. So the one-sided Z-transform process can be written as

$$F(z) = \mathcal{Z}[f(n)] = \sum_{n=0}^{\infty} f(n)z^{-n}. \tag{5.9}$$

So what exactly is $z$? Like the Laplace parameter $s$, it is a form of generalized frequency – specifically, $z$ is a single complex value of $e^{sT_0}$, which is often written as $e^s$ when a sample period $T_0$ of one unit is assumed. It is important to realize that $z$ is not a basis function, just as $s = \sigma + i\omega$ is not a basis function; to turn these single complex values into basis functions, you must raise $z$ to the power of $n$, in which $n$ is the series of integers from zero to $\infty$. Since $z = e^s$, raising $z$ to the power of $n$ produces the function $z^n = e^{sn}$, which is an exponentially decaying or growing sinusoidal basis function for each value of $s$, just as $e^{st}$ is an exponentially decaying or growing sinusoidal basis function in the case of the Laplace transform.

And what does the $z$-domain function $F(z)$ tell you? Just as the $s$-domain function $F(s)$ produced by the Laplace transform is a measure of the amount of each complex-exponential basis function $e^{st}$ that makes up the time-domain function $f(t)$, $F(z)$ tells you the amount of each basis function $z^n$ present in the sampled time-domain function $f(n)$.

Here is a little table showing the relevant variables, basis functions, transformation operators, and complex planes of the Laplace transform and the Z-transform:

| | |
|---|---|
| Complexfrequency − domainvariable | $s$ or $z$ |
| Time − domainvariable | $t$ or $n$ |
| Basisfunctions | $e^{-st}$ or $z^{-n}$ |
| Transformoperator | $\mathcal{L}$ or $\mathcal{Z}$ |
| Graphicalplane | $s$ −plane or $z$ −plane. |

The relationship between the Laplace-transform $s$-plane and the Z-transform $z$-plane is explained in the next subsection.

## The $z$-Plane

It is important to realize that the Z-transform $F(z)$ of the sampled time-domain function $f(n)$ is periodic in angular frequency $\omega$. That is, the values of $F(z)$ over any angular frequency range of $2\pi/T_0$ rad/sec are the same as the values over the next (or previous) range of $2\pi/T_0$. For example, $F(z)$ over the range of $\omega = 0$ to $2\pi/T_0$ rad/sec is identical to $F(z)$ over the range of $-2\pi/T_0$ to $0$ rad/sec and from $2\pi/T_0$ to $4\pi/T_0$ rad/sec, and so forth.

To see why that is true, let $T_0 = 1$ sec and allow $\omega$ to become $\omega + 2\pi$ in Eq. 5.9. If $F_{orig}(z)$ is the Z-transform of $f(n)$ at angular frequency $\omega$ and $F_{new}(z)$ is the Z-transform of $f(n)$ at angular frequency $\omega + 2\pi$, then

$$F_{orig}(z) = \sum_{n=0}^{\infty} f(n)z^{-n} = \sum_{n=0}^{\infty} f(n)e^{-n[\sigma+i\omega]} \tag{5.10}$$

and

$$F_{new}(z) = \sum_{n=0}^{\infty} f(n)e^{-n[\sigma+i(\omega+2\pi)]} = \sum_{n=0}^{\infty} f(n)e^{-n\sigma}e^{-in\omega}e^{-in2\pi}. \tag{5.11}$$

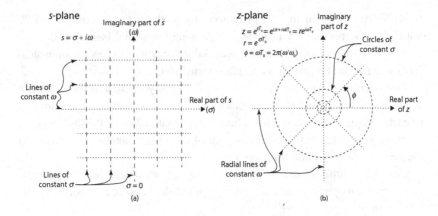

Figure 5.1 Lines of constant $\sigma$ and $\omega$ in (a) the $s$-plane and (b) the $z$-plane.

But the factor $e^{-in2\pi}$ is unity for all integer values of $n$:

$$e^{-in2\pi} = \cos(n2\pi) - i\sin(n2\pi) = 1 - i(0) = 1 \qquad (5.12)$$

so Eq. 5.11 is

$$F_{new}(z) = \sum_{n=0}^{\infty} f(n)e^{-n\sigma}e^{-in\omega}(1) = \sum_{n=0}^{\infty} f(n)z^{-n} = F_{orig}(z). \qquad (5.13)$$

This periodicity of $F(z)$ with $\omega$ suggests that polar rather than rectilinear coordinates may be useful in displaying the Z-transform of sampled time-domain functions. You can see how that works in Fig. 5.1, which shows both the $s$-plane and the $z$-plane. As indicated in this figure, the vertical lines of constant $\sigma$ in the $s$-plane can be mapped onto concentric circles of radius $r = e^{\sigma T_0}$ in the $z$-plane, while the horizontal lines of constant $\omega$ in the $s$-plane are mapped onto radial lines in the $z$-plane.

You should note that the angle that the radial constant-$\omega$ lines make with the positive real axis is given by $\omega T_0$, and the sample period $T_0$ can be written in terms of the sampling angular frequency $\omega_0$ as $T_0 = 2\pi/\omega_0$. So the angle $\phi$ of the radial lines in Fig. 5.1b is

$$\phi = \omega T_0 = \omega\left(\frac{2\pi}{\omega_0}\right) = 2\pi\frac{\omega}{\omega_0}, \qquad (5.14)$$

which means that the angle is simply the ratio of the angular frequency to the sampling angular frequency (that is, $\omega/\omega_0$) converted to radians by multiplying

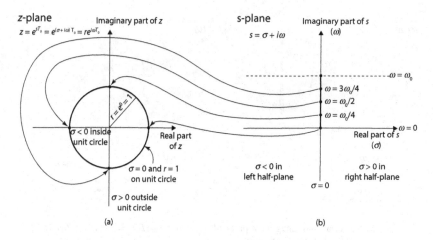

Figure 5.2 Mapping the $\sigma = 0$ axis from the $s$-plane onto the unit circle in the $z$-plane.

by $2\pi$. This explains why you may see the angle in the $z$-plane labeled as $\omega$ in some texts and websites – the authors of those books and sites have assumed a sampling period $T_0$ of unity, which means that $2\pi/\omega_0 = 1$, so $\phi = \omega$.

Additional details of the mapping from the $s$-plane to the $z$-plane are shown in Fig. 5.2. This figure shows how four points on the $\sigma = 0$ axis in the $s$-plane map onto the unit circle (that is, the circle with radius $r = e^0 = 1$) in the $z$-plane. Those four points have angular frequency values of $\omega = 0$, $\omega_0/4$, $\omega_0/2$, and $3\omega_0/4$; the point at $\omega = \omega_0$ maps onto the same point in the $Z$-plane as $\omega = 0$.

As indicated in this figure, all points with $\sigma < 0$ (that is, the entire left half of the $s$-plane) map into the inside of the unit circle in the $z$-plane, while all points with $\sigma > 0$ (the entire right half of the $s$-plane) map onto the outside of the unit circle in the $z$-plane. This is important when you consider the region of convergence (ROC) of the Z-transform, as you'll see in the Z-transform examples in Section 5.2.

Before delving into the inverse Z-transform, you may find it instructive to consider the three-dimensional $z$-plane plot shown in Fig. 5.3. In figures such as this, the real ($\text{Re}(z)$) and imaginary ($\text{Im}(z)$) parts of $z = e^{\sigma+i\omega}$ are plotted along the $x$- and $y$-axes, while the complex $z$-domain function $F(z)$ is plotted on the $z$-axis. As in the case of the Laplace-transform $s$-plane function $F(s)$, the complex nature of the generalized frequency-domain function $F(z)$ means that two plots are needed to convey all information contained in the function. Those two plots are usually the real and imaginary parts of the

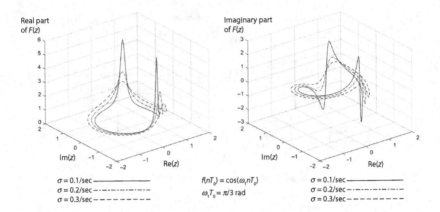

Figure 5.3 Real and imaginary parts of $F(z)$ for $f(nT_0) = \cos(\omega_1 nT_0)$ for $\omega_1 T_0 = \pi/3$ rad at three values of $\sigma$.

function, as in this figure, but you may also encounter three-dimensional plots of the magnitude and phase of $F(z)$.

As you may have anticipated if you've worked through the examples in Chapter 2, the $z$-plane function $F(z)$ shown in Fig. 5.3 is the result of taking the Z-transform of a positive-time cosine function (that is, a cosine function that has zero amplitude for all samples before time $t = 0$). Specifically, the sampled time-domain function in this case is $f(nT_0) = 0$ for all $n < 0$ and $f(nT_0) = \cos(\omega_1 nT_0)$ for $n \geq 0$, which is one of the examples you can read about later in this chapter. As you will see, this Z-transform has a ROC of $|z| = |e^\sigma e^{i\omega}| > 1$, which means that any positive value of $\sigma$ will cause the Z-transform series to converge for this sequence.

In Fig. 5.3, the angular frequency $\omega_1$ times the sample period $T_0$ is $\pi/3$ rad; this tells you the relationship between the angular frequency $\omega_1$ of the cosine function and the sampling angular frequency $\omega_0$. To find $\omega_1$ in terms of $\omega_0$, use Eq. 5.14 with $\omega_1$ in place of $\omega$:

$$\phi = \omega_1 T_0 = \omega_1 \left(\frac{2\pi}{\omega_0}\right) = 2\pi\left(\frac{\omega_1}{\omega_0}\right) = \frac{\pi}{3}, \qquad (5.15)$$

which means

$$\omega_1 = \left(\frac{1}{2\pi}\right)\omega_0\left(\frac{\pi}{3}\right) = \frac{\omega_0}{6}. \qquad (5.16)$$

There are three curves in both Fig. 5.3a and Fig. 5.3b, corresponding to $\sigma$ values of 0.1/sec, 0.2/sec, and 0.3/sec. That means that the radii of the constant-$\sigma$ circles of these three curves are $e^{0.1} = 1.11$, $e^{0.2} = 1.22$, and

$e^{0.3} = 1.35$ units. So these curves all lie outside the unit circle, within the ROC for this function. And what do those curves tell you?

As mentioned above, the $z$-domain function $F(z)$ is a measure of the amount of each $z^n$ basis function that makes up the corresponding time-domain function – in this case, the sampled time-domain function $f(nT_0) = \cos(\omega_1 nT_0)$. Weighting those basis functions by the $z$-domain function $F(z)$ and combining them by integration synthesizes the time-domain function $f(nT_0)$. That synthesis process is accomplished using the inverse Z-transform, which is the subject of the next subsection.

## The Inverse Z-Transform

The mathematical statement of the inverse Z-transform looks like this:

$$f(nT_0) = \frac{1}{2\pi i} \oint_C F(z)z^{n-1}dz, \qquad (5.17)$$

in which the integral is performed over a closed path (C) within the ROC surrounding the origin in the $z$-plane. As in the case of the inverse Laplace transform, this equation is rarely used to find the inverse Z-transform in practice; for most applications the inverse Z-transform can be determined by finding $F(z)$ and the corresponding sampled time-domain function $f(n)$ in a table of Z-transforms (perhaps after using partial-fraction or series methods to express $F(z)$ in terms of basic functions).

You may find it helpful to consider the relationship of the inverse Z-transform expressed by Eq. 5.17 to the inverse discrete-time Fourier transform. To do that, start by recalling that on the unit circle in the $z$-plane, $\sigma = 0$, so $z = e^{(\sigma+i\omega)T_0} = e^{i\omega T_0}$. And since the $s$-plane $\omega$ axis maps onto the $z$-plane unit circle, integrating around the unit circle in the $z$-plane is equivalent to integrating along the $\sigma = 0$ line from $\omega = -\pi$ to $\omega = +\pi$ in the $s$-plane. Also note that the increment $dz$ is related to the angular-frequency increment $d\omega$:

$$\frac{dz}{d\omega} = \frac{de^{i\omega T_0}}{d\omega} = iT_0 e^{i\omega T_0} = iT_0 z \qquad (5.18)$$

so $dz = iT_0 z \, d\omega$. Hence Eq. 5.17 can be written as

$$f(nT_0) = \frac{1}{2\pi i} \oint_C F(z)z^{n-1}dz = \frac{1}{2\pi i} \int_{-\pi}^{\pi} F(z)z^{n-1}(iT_0)z \, d\omega$$

$$= \frac{T_0}{2\pi} \int_{-\pi}^{\pi} F(z)z^n d\omega, \qquad (5.19)$$

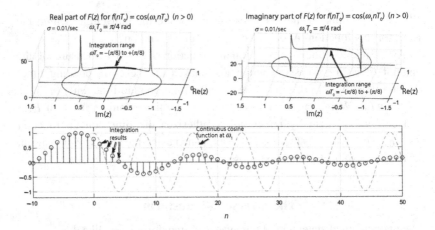

Figure 5.4  Inverse Z-transform of $F(z)$ for $f(nT_0) = \cos(\omega_1 nT_0)$ with $\omega_1 T_0 = \pi/4$ and integration range of $\omega T_0 = -\pi/8$ rad to $\omega T_0 = +\pi/8$ rad.

which is the inverse Fourier transform for a discrete-time sequence with sampling period $T_0$. So as long as the unit circle in the $z$-plane lies within the ROC, the inverse Z-transform with $\sigma = 0$ and the inverse discrete-time Fourier transform will produce the same result.

And exactly how does the inverse Z-transform of the $z$-domain function $F(z)$ give the discrete-time function $f(nT_0)$? The process is the discrete-time equivalent of the synthesis of continuous-time functions $f(t)$ performed by the inverse Laplace transform.

You can see that process in action for the inverse Z-transform of $f(nT_0) = \cos(\omega_1 nT_0)$ for $n > 0$ in Figs. 5.4 through 5.6 (as mentioned above, cosine functions are one of the examples of the Z-transform described in the next section of this chapter, where you can see how this result comes about). The angular frequency of the positive-time cosine function is chosen to be $\omega_1 = \pi/4$ rad/sec with $\sigma = 0.01$/sec in each of these figures, but the range of integration is varied to show the contributions of low, medium, and high frequencies (relative to the sampling angular frequency $\omega_0$) to the synthesis of $f(nT_0)$.

In Fig. 5.4, the range of integration, indicated by the dark portions of the $F(z)$ curves, extends from $\omega T_0 = -\pi/8$ rad to $\omega T_0 = +\pi/8$ rad, so only the lowest angular frequencies are included. The results of integrating the product $F(z)$ and $z^n$ are shown in the lower portion of this figure; these results are normalized to a value of unity to make it easier to compare them to the cosine function with angular frequency $\omega_1$ (shown as a dashed line in this figure).

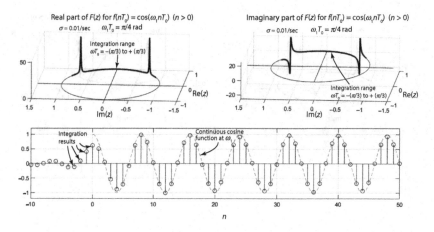

Figure 5.5 Inverse Z-transform of $F(z)$ for $f(nT_0) = \cos(\omega_1 nT_0)$ with $\omega_1 T_0 = \pi/4$ and integration range of $\omega T_0 = -\pi/3$ rad to $\omega T_0 = +\pi/3$ rad.

As you can see, the dominant frequency of the integrated results is considerably lower than $\omega_1$, and the amplitude of the oscillations rolls off with increasing positive $n$. Additionally, the synthesized samples of $f(nT_0)$ for $n < 0$ have significant amplitude rather than the desired zero amplitude for negative time. As described in the discussions of the Laplace transform in chapters 1 and 2, additional frequency components with appropriate phase (given by the mixture of cosine and sine functions) are needed to cancel the contributions of these low-frequency components for negative-time samples.

The effect of including additional frequency components is shown in Fig. 5.5, for which the range of integration extends from $\omega T_0 = -\pi/3$ rad to $\omega T_0 = +\pi/3$ rad. In this case, the large-amplitude peaks in the real portion of $F(z)$ and the sign-flipping peaks in the imaginary portion of $F(z)$ are included in the integration (again indicated by the dark portion of the curve in the upper portions of the figure), and the impact of those peaks on the integrated result is significant.

You can see that impact in the lower portion of Fig. 5.5; the result of integrating over this range of angular frequencies is much closer to a cosine function for $n > 0$ than the result of using a narrower integration range, and the negative-time ($n < 0$) samples have been driven considerably closer to zero. The same rationale described in Section 2.3 for continuous-time sinusoidal functions applies to this discrete-time function and, just as in the continuous case, additional high-frequency components are required to further reduce the

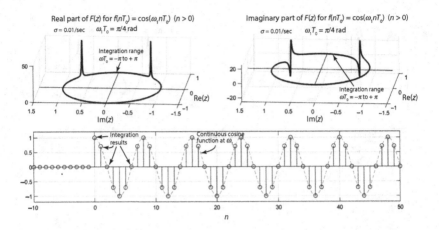

Figure 5.6 Inverse Z-transform of $F(z)$ for $f(nT_0) = \cos(\omega_1 nT_0)$ with $\omega_1 T_0 = \pi/4$ and integration range of $\omega T_0 = -\pi$ rad to $\omega T_0 = +\pi$ rad.

deviation of the negative-time samples from zero and the slight differences in the positive-time samples from a cosine function.

Those higher-frequency components are included in the integration results shown in Fig. 5.6, for which the range of integration covers the entire range from $\omega T_0 = -\pi$ rad to $\omega T_0 = +\pi$ rad. As expected, integrating over a full period of $2\pi$ radians in $\omega T_0$ produces a discrete-time result that is indistinguishable from the sampled time-domain function $f(nT_0)$ of which $F(z)$ is the unilateral Z-transform.

With this explanation of the meaning of the Z-transform and the function $F(z)$ in hand, you should be ready to work through the Z-transform process of several basic functions. That is the subject of the next section.

## 5.2  Examples of the Z-transform

This section shows the Z-transform of six basic discrete time-domain functions, including the unit-impulse and unit-step functions, exponential functions, and sinusoidal functions. Like the Laplace transform, the Z-transform is a linear operation with characteristics that allow the transforms of basic functions to be combined to produce the transform of more-complicated functions. This property and several other useful characteristics of the Z-transform are described in Section 5.3.

For each of the examples in this section, the sample period $T_0$ is taken as unity, and $T_0$ is not explicitly written. That means you will see values for quantities such as $\sigma = 0.01$ and $\omega = \pi/2$ rad written without the usual SI units of seconds in the denominator, since these values represent $\sigma T_0$ and $\omega T_0$ and the time units of $T_0$ have canceled the "per time" units of the frequencies $\sigma$ and $\omega$.

### The Unit-Impulse Function

The unit-impulse function $\delta(n)$ is the discrete-time version of the continuous-time delta function $\delta(t)$ described in Section 1.3. Just as $\delta(t)$ has nonzero value only at time $t = 0$, $\delta(n)$ has nonzero value only when its argument ($n$ in this case) is zero. So

$$\delta(n) = \begin{cases} 1, & (n = 0) \\ 0, & (n \neq 0) \end{cases}, \tag{5.20}$$

in which $n$ represents the sample number. Note that, unlike the continuous-time delta function, which must become infinitely tall to maintain an area of unity as its width shrinks to zero, the amplitude of the unit-impulse function is defined as one for a single sample. Hence multiplying a continuous-time function $f(t)$ by $\delta(n)$ produces a single nonzero sample with amplitude equal to the amplitude of $f(t)$ at time $t = 0$.

Note also that it is the argument of the unit-impulse function that determines the number of the nonzero sample. So $\delta(n - 5)$ has value of unity when $n - 5 = 0$, which means $n = 5$. Multiplying $f(t)$ by $\delta(n - 5)$ gives the single value $f(t)\delta(n - 5) = f(5)$.

To determine the Z-transform of the unit-impulse function, set

$$f(n) = \delta(n) \tag{5.21}$$

in the equation for the unilateral Z-transform (Eq. 5.9):

$$F(z) = \sum_{n=0}^{\infty} f(n)z^{-n} = \sum_{n=0}^{\infty} \delta(n)z^{-n}. \tag{5.22}$$

But $\delta(n) = 0$ unless $n = 0$, so only one term contributes to the summation, and the z-domain representation of the unit-impulse function is

$$F(z) = \sum_{n=0}^{\infty} \delta(n)z^{-n} = \delta(0)z^0 = (1)(1) = 1. \tag{5.23}$$

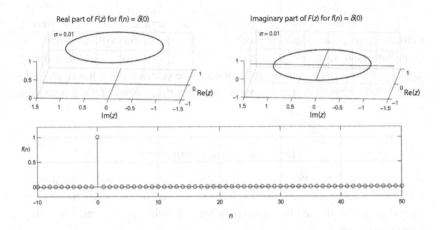

Figure 5.7  $F(z)$ and inverse Z-transform for the sequence $f(n) = \delta(0)$.

Thus the Z-transform output $F(z)$ is a constant, real function with unity amplitude for the sequence $f(n) = \delta(n)$. The summation converges for any value of $z$, so the ROC is the entire $z$-plane in this case.

This result shouldn't be surprising if you have worked through the Fourier- and Laplace-transform discussions in chapters 1 and 2, because $F(z) = 1$ means that at any value of $\sigma$, all angular frequencies ($\omega$), weighted equally, combine to produce a time-domain delta function.

You can see the real (unity) and imaginary (zero) parts of $F(z)$ for $\sigma = 0.01$ in the upper portion of the three-dimensional $z$-plane plot in Fig. 5.7. As explained above, the unitless value of 0.01 represents the quantity $\sigma T_0$ with sampling period $T_0$ set to unity.

And do these values of $F(z)$ produce the time-domain unit-impulse function when inserted into the inverse Z-transform equation (Eq. 5.19)? You can see that they do in the lower portion of this figure, which is the output of the inverse Z-transform integrated around the full circle from $\omega = -\pi$ to $\omega = \pi$ with radius $e^{0.01} = 1.1$. As expected, the time-domain result is a single non-zero sample at $n = 0$.

## The Unit-Step Function

The discrete-time unit-step function is the sampled version of the positive-time constant function $f(t) = c$ for $t \geq 0$ discussed in Section 2.1. In this case, the sampled function has a value of unity for all samples $n \geq 0$ and zero for all samples with negative $n$:

$$u(n) = \begin{cases} 1, & n \geq 0 \\ 0, & n < 0 \end{cases}, \tag{5.24}$$

in which $u(n)$ represents the discrete-time unit-step function.

To find the Z-transform of the unit-step function, set the time-domain function $f(n)$ equal to $u(n)$

$$f(n) = u(n) \tag{5.25}$$

and insert this into the Z-transform equation (Eq. 5.9):

$$F(z) = \sum_{n=0}^{\infty} f(n)z^{-n} = \sum_{n=0}^{\infty} u(n)z^{-n} = \sum_{n=0}^{\infty} \left(\frac{1}{z}\right)^n. \tag{5.26}$$

The terms of this summation form a power series in $1/z$, and as long as $|x| < 1$, that power series converges:

$$\sum_{n=0}^{\infty} x^n = \frac{1}{1-x}. \tag{5.27}$$

Substituting $1/z$ for $x$ in this relation gives

$$F(z) = \sum_{n=0}^{\infty} \left(\frac{1}{z}\right)^n = \frac{1}{1-\frac{1}{z}} = \frac{z}{z-1}, \tag{5.28}$$

in which the condition for convergence is that $|z| > 1$.

The real and imaginary parts of $F(z)$ for $\sigma = 0.01$ are shown in the upper portion of the three-dimensional $z$-plane plot in Fig. 5.8. Once again, the value of 0.01 represents the quantity $\sigma T_0$ with sampling period $T_0$ set to unity.

Comparing the $z$-domain function $F(z)$ for the discrete-time unit-step function to the $s$-domain function $F(s)$ for the continuous-time step function reveals that in both cases the real part of the generalized frequency-domain function has a single, symmetrical peak about $\omega = 0$ while the imaginary part has a sign reversal at that frequency. The reasons for this behavior are explained in Section 2.1 in terms of the sine and cosine components that make up the time-domain functions $f(t)$ and $f(n)$.

The result of taking the inverse Z-transform of $F(z)$ is shown in the lower portion of this figure, for which the integration has been performed around the full circle from $\omega = -\pi$ to $\omega = \pi$ with radius $e^{0.01} = 1.1$. The resulting time-domain function has zero amplitude for all negative-$n$ samples and unity amplitude for $n$ equal to or greater than zero.

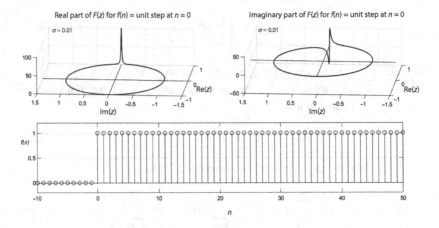

Figure 5.8 $F(z)$ and inverse Z-transform for $f(n)$ = unit-step function.

## Constant-to-a-Power Functions

A discrete-time function often encountered in Z-transform applications is a sequence of values given by a constant (represented by $a$ in this section) raised to an integer power corresponding to the sample number $n$:

$$f(n) = a^n \qquad n \geq 0 \tag{5.29}$$

for positive values of $n$.

Note that this type of function differs from the power of $t$ functions discussed in Section 2.4 – in that case the base $(t)$ increased while the exponent $(n)$ remained constant, but in this case the base $(a)$ remains constant while the exponent $(n)$ increases.

The unilateral Z-transform of the function defined in Eq. 5.29 is given by

$$F(z) = \sum_{n=0}^{\infty} f(n)z^{-n} = \sum_{n=0}^{\infty} a^n z^{-n} = \sum_{n=0}^{\infty} \left(\frac{a}{z}\right)^n, \tag{5.30}$$

and using the same power-series relation (Eq. 5.27) as in the previous example makes this

$$F(z) = \sum_{n=0}^{\infty} \left(\frac{a}{z}\right)^n = \frac{1}{1 - \frac{a}{z}} = \frac{z}{z - a}, \tag{5.31}$$

and in this case the convergence requirement is that $|z| > |a|$. Hence the ROC consists of all points in the $z$-plane outside the circle with radius equal to $e^{\sigma} = a$.

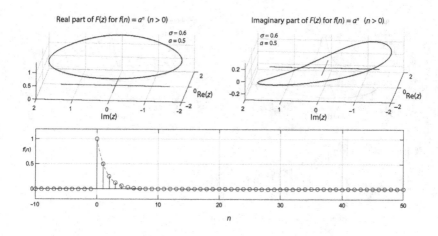

Figure 5.9  $F(z)$ and inverse Z-transform for $f(n) = a^n$.

As you can see in the bottom portion of Fig. 5.9, for $\sigma = 0.6$ and $a = 0.5$, the amplitude of the time-domain sequence $f(n)$ decreases rapidly toward zero with increasing $n$, and the real and imaginary parts of $F(z)$ shown in the top portion are similar to those of the Z-transform of the unit-impulse function. That shouldn't be surprising, since in both cases a wide range of frequencies is needed to produce a narrow time-domain response. However, since the unit-impulse response is symmetric about $\omega = 0$, in that case only the real portions of $F(z)$ (that is, the cosine components) have nonzero amplitude, but the asymmetric nature of the $a^n$ sequence (for positive $n$ only) requires contributions from the imaginary portions (sine components) of $F(z)$. The sign of those components changes at $\omega = 0$, as needed to cancel the contributions of the cosine components for negative values of $n$.

### Exponential Functions

Although exponential functions such as

$$f(n) = e^{-an} \qquad n \geq 0 \tag{5.32}$$

can be considered to be versions of the "constant-to-a-power" functions discussed in the previous example, the frequent appearance of exponential sequences in Z-transform applications makes it worth a bit of your time to consider the nature of the z-domain function $F(z)$ for such functions.

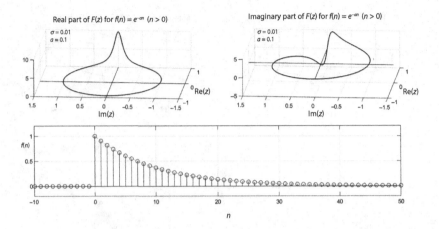

Figure 5.10  $F(z)$ and inverse Z-transform for $f(n) = e^{-an}$.

The unilateral Z-transform for the exponential sequence defined in Eq. 5.32 is

$$F(z) = \sum_{n=0}^{\infty} f(n)z^{-n} = \sum_{n=0}^{\infty} e^{-an}z^{-n} = \sum_{n=0}^{\infty} \left(e^a z\right)^{-n}, \qquad (5.33)$$

in which $a$ represents the frequency (that is, the "rate of $1/e$ steps") of the sampled time-domain function $f(n)$.

Once again, the same power-series relation (Eq. 5.27) used in the two previous examples can be used to make this

$$F(z) = \sum_{n=0}^{\infty} \left(e^a z\right)^{-n} = \frac{1}{1 - \frac{1}{ze^a}} = \frac{z}{z - e^{-a}} \qquad (5.34)$$

as long as $|ze^a| > 1$.

The upper portion of Fig. 5.10 shows the real and imaginary parts of the output $F(z)$ of the Z-transform of an exponential function of this type with $a = 0.1$ and $\sigma = 0.01$. For these parameters the shape of $F(z)$ is between that of the unit-step function (with constant amplitude for positive values of $n$) and the constant-to-a-power function (with rapidly decreasing amplitude for positive $n$). For larger values of $a$, the time-domain sequence $f(n)$ will decrease more quickly with increasing values of $n$, and the width of the corresponding function $F(z)$ will increase as higher frequencies are needed to produce the narrower time-domain function.

## Sinusoidal Functions

The final two examples in this section are the sampled sinusoidal time-domain functions of the form $\cos(\omega n)$ and $\sin(\omega n)$. Just as in the case of the Laplace transforms of continuous-time sinusoidal functions discussed in Section 2.3 the Z-transforms of these functions are a bit more complicated than those of unit-step, constant, and exponential functions. Happily, several of the mathematical techniques that proved useful in determining the Laplace transform of sinusoidal functions are also helpful for Z-transforms, and the conceptual understanding of the shape of the real and imaginary parts of $F(s)$ (that is, the relative amounts of sine and cosine components) can also be applied to $F(z)$.

A discrete-time cosine function with angular frequency $\omega_1$ for positive values of $n$ can be written as

$$f(n) = \cos(\omega_1 n) \qquad n \geq 0 \qquad (5.35)$$

and plugging this expression for $f(n)$ into the definition of the Z-transform (Eq. 5.9) gives

$$F(z) = \sum_{n=0}^{\infty} f(n)z^{-n} = \sum_{n=0}^{\infty} \cos(\omega_1 n)z^{-n}. \qquad (5.36)$$

Now use the inverse Euler relation for the cosine function (Eq. 1.5 in Section 1.1) to express the cosine function as the sum of two exponentials:

$$\cos(\omega_1 n) = \frac{e^{i\omega_1 n} + e^{-i\omega_1 n}}{2}.$$

This makes Eq. 5.36 look like this:

$$F(z) = \sum_{n=0}^{\infty} \cos(\omega_1)z^{-n} = \sum_{n=0}^{\infty} \left( \frac{e^{i\omega_1 n} + e^{-i\omega_1 n}}{2} \right) z^{-n} \qquad (5.37)$$

$$= \frac{1}{2} \sum_{n=0}^{\infty} \left[ \left( ze^{-i\omega_1} \right)^{-n} + \left( ze^{i\omega_1} \right)^{-n} \right] \qquad (5.38)$$

$$= \frac{1}{2} \sum_{n=0}^{\infty} \left( \frac{1}{ze^{-i\omega_1}} \right)^n + \frac{1}{2} \sum_{n=0}^{\infty} \left( \frac{1}{ze^{i\omega_1}} \right)^n. \qquad (5.39)$$

The next step is to use the power-series relation given by Eq. 5.27 to make this

$$F(z) = \frac{1}{2}\left(\frac{1}{1 - \frac{1}{ze^{-i\omega_1}}}\right) + \frac{1}{2}\left(\frac{1}{1 - \frac{1}{ze^{i\omega_1}}}\right)$$

$$= \frac{1}{2}\left(\frac{ze^{-i\omega_1}}{ze^{-i\omega_1} - 1}\right) + \frac{1}{2}\left(\frac{ze^{i\omega_1}}{ze^{i\omega_1} - 1}\right)$$

as long as $|1/ze^{-i\omega_1}| < 1$ and $|1/ze^{i\omega_1}| < 1$, which means $|z| > 1$.

These two exponential terms can be added after finding their common denominator:

$$F(z) = \frac{1}{2}\left[\frac{\left(ze^{-i\omega_1}\right)\left(ze^{i\omega_1} - 1\right)}{\left(ze^{-i\omega_1} - 1\right)\left(ze^{i\omega_1} - 1\right)} + \frac{\left(ze^{i\omega_1}\right)\left(ze^{-i\omega_1} - 1\right)}{\left(ze^{i\omega_1} - 1\right)\left(ze^{-i\omega_1} - 1\right)}\right]$$

$$= \frac{1}{2}\left[\frac{z^2 - ze^{-i\omega_1} + z^2 - ze^{i\omega_1}}{z^2 - ze^{-i\omega_1} - ze^{i\omega_1} + 1}\right] = \frac{1}{2}\left[\frac{2z^2 - z\left(e^{i\omega_1} + e^{-i\omega_1}\right)}{z^2 - z\left(e^{i\omega_1} + e^{-i\omega_1}\right) + 1}\right].$$

The final step in determining $F(z)$ is to use the inverse Euler relation to convert the exponential terms into $2\cos(\omega_1)$, giving

$$F(z) = \frac{z^2 - z\cos(\omega_1)}{z^2 - 2z\cos(\omega_1) + 1}. \tag{5.40}$$

This is the unilateral Z-transform of the discrete-time cosine function in the ROC $|z| > 1$.

The real and imaginary parts of $F(z)$ are shown in the upper portion of Fig. 5.11, and they should look familiar if you've worked through the first section of this chapter and the discussion of the Laplace transform $F(s)$ of the continuous-time cosine function in Chapter 2. In this plot, the angular frequency $\omega_1$ is chosen to be $\pi/10$, and $\sigma = 0.01$. The real part of $F(z)$ has two positive peaks, one at $+\omega_1$ and one at $-\omega_1$; the angle of those peaks from the real axis in the $z$-plane can be found using Eq. 5.14 with $\omega = \omega_1 = \pi/10$ and $T_0 = 1$ (which makes $\omega_0 = 2\pi$):

$$\phi = \omega_1 T_0 = 2\pi\frac{\omega_1}{\omega_0} = 2\pi\frac{\pi/10}{2\pi} = \pi/10 \tag{5.41}$$

or about $18°$.

As explained in Chapter 2, the sine components in the Laplace or Z-transform of a positive-time cosine function are required to produce zero amplitude for negative time ($t$) in the continuous-time case and negative sample number ($n$) in the discrete-time case. You can see the result of

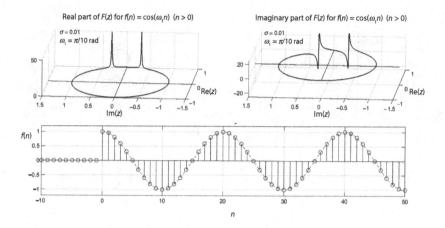

Figure 5.11 $F(z)$ and inverse Z-transform for $f(n) = \cos(\omega_1 n)$.

taking the inverse Z-transform of $F(z)$ in the lower portion of Fig. 5.11; the integration has been performed over the full range of angular frequency from $\omega = -\pi$ to $\omega = +\pi$ to synthesize the sampled time-domain function $f(n)$.

The same steps shown above for finding the Z-transform of the discrete-time cosine function can be used for the discrete-time sine function. That function can be written as

$$f(n) = \sin(\omega_1 n) \qquad n \geq 0 \tag{5.42}$$

and its Z-transform is

$$F(z) = \frac{z \sin(\omega_1)}{z^2 - 2z \cos(\omega_1) + 1} \tag{5.43}$$

in the ROC $|z| > 1$ (you can see the steps leading to this in one of the chapter-end problems and its online solution).

The real and imaginary parts of $F(z)$ are shown in the upper portion of Fig. 5.12, once again with angular frequency $\omega_1$ of $\pi/10$ and $\sigma = 0.01$. The same rationale given in Chapter 2 for the shape of real and imaginary parts of $F(s)$ of the continuous-time sine function applies to $F(z)$ for the discrete-time version, and the result of taking the inverse Z-transform of $F(z)$ is shown in the lower portion of this figure.

Like the Laplace transform, the Z-transform has several useful characteristics that allow you to use the basic examples presented in this chapter to find the forward and inverse transform of more complicated functions. You can read about the most useful of those characteristics in the next section.

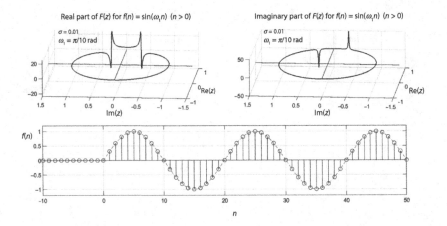

Figure 5.12  $F(z)$ and inverse Z-transform for $f(n) = \sin(\omega_1 n)$.

## 5.3  Characteristics of the Z-transform

The Z-transform shares many of the useful characteristics of the Laplace transform, and the purpose of this section is to describe several of those characteristics, including linearity, time-shifting, multiplication by exponentials, differentiation, convolution, and the initial- and final-value theorems. The comprehensive texts listed in the bibliography discuss additional properties of the Z-transform that you may find useful in certain applications, but the characteristics presented in this section are among the most widely applicable in determining the Z-transform of more-complicated discrete-time functions by combining the transforms of basic functions.

### Linearity

As explained in Section 3.1, linearity encompasses two properties: homogeneity and additivity. Like the Laplace transform, the Z-transform is a linear process, so it is both homogeneous:

$$\mathcal{Z}[cf(n)] = c\mathcal{Z}[f(n)] = cF(z), \tag{5.44}$$

in which $c$ represents a constant, and additive:

$$\mathcal{Z}[f(n) + g(n)] = \mathcal{Z}[f(n)] + \mathcal{Z}[g(n)] = F(z) + G(z), \tag{5.45}$$

in which $F(z)$ and $G(z)$ represent the Z-transforms of $f(n)$ and $g(n)$, respectively.

To see why the Z-transform is a linear process, start by producing a combined function by adding two scaled functions: $f(n)$ scaled by constant $c_1$ and $g(n)$ scaled by constant $c_2$. Then insert the sum $c_1 f(n) + c_2 g(n)$ into the definition of the Z-transform (Eq. 5.9):

$$\mathcal{Z}[c_1 f(n) + c_2 g(n)] = \sum_{n=0}^{\infty} [c_1 f(n) + c_2 g(n)] z^{-n}$$

$$= \sum_{n=0}^{\infty} [c_1 f(n) z^{-n} + c_2 g(n) z^{-n}]$$

$$= c_1 \sum_{n=0}^{\infty} f(n) z^{-n} + c_2 \sum_{n=0}^{\infty} g(n) z^{-n}$$

$$= c_1 F(z) + c_2 G(z). \tag{5.46}$$

So the scaling and the summing can be done either in the discrete-time domain or in the $z$-domain, and the result will be the same.

The rationale for the linearity of the Z-transform (beyond the mathematical justification above) parallels that of the Laplace transform presented in Chapter 3. That is, scaling and adding discrete-time functions such as $f(n)$ and $g(n)$ has the same effect as scaling and adding the basis functions that make up those functions. That effect can also be achieved by scaling and adding the Z-transforms $F(z)$ and $G(z)$, which tell you the "amounts" of basis functions that make up $f(n)$ and $g(n)$.

## Time-Shifting

One of the most useful properties of the Z-transform in signal processing concerns the Z-transform of a discrete-time function that has been shifted in time. The "right-shift" version of this property says that the Z-transform of a function shifted toward later times by $n_1$ samples can be found by multiplying the Z-transform $F(z)$ of the unshifted sequence by the factor $z^{-n_1}$.

There is also a "left-shift"" version of the time-shift property in which the Z-transform of the unshifted sequence is multiplied by $z^{+n_1}$, but in that case it's also necessary to subtract off the contributions of the samples that are left-shifted past $n = 0$. The mathematical statements of both the left-shift and the right-shift versions of the Z-transform time-shift property are presented below.

The time-shift property of the Z-transform has led to the widespread use of the notation "$z^{-1}$" to represent a unit-delay element in a discrete-time circuit.

The right-shift version of the Z-transform time-shift property can be written as

$$\mathcal{Z}[f(n - n_1)] = z^{-n_1} \mathcal{Z}[f(n)] = z^{-n_1} F(z), \qquad (5.47)$$

in which the integer constant $n_1$ in the exponent of the multiplying factor $z^{-n_1}$ determines the number of samples by which the discrete-time function $f(n)$ is shifted toward later time.

You can see the mathematical basis for this property by considering a causal sequence such as $f(n)$. Recall that for a causal sequence, all samples with negative values of $n$ (that is, to the left of sample with $n = 0$) must have zero amplitude. So shifting this sequence toward higher positive values of $n$ by $n_1$ samples means that the first $n_1$ samples of the shifted sequence $f(n - n_1)$ must have zero amplitude, and the Z-transform of the shifted sequence looks like this:

$$\mathcal{Z}[f(n - n_1)] = \sum_{n=0}^{\infty} f(n - n_1)z^{-n} = \sum_{n=n_1}^{\infty} f(n - n_1)z^{-n}. \qquad (5.48)$$

The second equality is true because the first $n_1$ samples of the sequence $f(n - n_1)$ are zero, so starting the summation at $n = n_1$ gives the same result as starting at $n = 0$.

Now use the change of variable $k = n - n_1$, which means $n = k + n_1$:

$$\mathcal{Z}[f(n - n_1)] = \sum_{k=0}^{\infty} f(k)z^{-(k+n_1)}$$

$$= \sum_{k=0}^{\infty} f(k)z^{-k}z^{-n_1} = z^{-n_1} F(z), \qquad (5.49)$$

in which $F(z)$ represents the Z-transform of the unshifted sequence.

For an intuitive understanding of why the process of multiplying a Z-transform such as $F(z)$ by $z^{-n_1}$ is equivalent to shifting $f(n)$ by $n_1$ samples toward later time, recall the explanation presented in Section 3.2 about the effect of multiplying $F(s)$ by the complex exponential factor $e^{-sa} = e^{-(\sigma+i\omega)a}$. That multiplication modifies the phase of each basis function by just the right amount for the mixture to synthesize the shifted time-domain function, and the same reasoning can be applied to the Z-transform. That is true because $z = e^{-s}$, so multiplying by $z^{-n_1}$ means multiplying by the complex exponential $e^{-sn_1}$.

Similar logic leads to the left-shift version of the time-shift property, which states that the Z-transform of the left-shifted sequence $f(n + n_1)$ is given by

$$\mathcal{Z}[f(n + n_1)] = z^{n_1} F(z) - \sum_{k=0}^{n_1-1} f(k)z^{n_1-k}, \qquad (5.50)$$

in which $n_1$ is again a positive integer and the terms in the summation are subtracted to remove the contributions of samples which have been left-shifted past $n = 0$. You can see details of the derivation of this equation and how it works in one of the chapter-end problems and its online solution.

## Multiplication by Exponentials

Multiplication of the discrete-time sequence $f(n)$ by a complex exponential sequence in the time domain results in scaling of the Z-transform output $F(z)$ in the $z$-domain. Specifically, if the multiplying sequence is

$$z_1^n = e^{(\sigma_1 + i\omega_1)n} \qquad (5.51)$$

then the Z-transform result is

$$\mathcal{Z}[z_1^n f(n)] = F\left(\frac{z}{z_1}\right). \qquad (5.52)$$

The mathematical justification of this property comes about by inserting the product $z_1^n f(n)$ into the Z-transform, which looks like this:

$$\mathcal{Z}[z_1^n f(n)] = \sum_{n=0}^{\infty} z_1^n f(n)z^{-n} = \sum_{n=0}^{\infty} f(n)z_1^n z^{-n}$$

$$= \sum_{n=0}^{\infty} f(n)\left(\frac{z}{z_1}\right)^{-n} = F\left(\frac{z}{z_1}\right).$$

So how does the $z$-domain function $F(z/z_1)$ compare to $F(z)$? This resembles the frequency-domain scaling discussed in Section 3.3, but since $z_1$ is a complex exponential, the effects of dividing the argument of the function $F(z)$ by $z_1$ depend on the nature of $z_1$. To understand those effects, consider what happens if the angular frequency $\omega_1$ in the time-domain multiplying factor $z_1^n = e^{(\sigma_1 + i\omega_1)n}$ is zero. In that case, dividing the argument of the $z$-domain function $F(z)$ by $z_1$ has the effect of scaling (expanding or contracting) the $z$-plane. One effect of that scaling is to cause the poles and zeros of the function $F(z)$ to move along radial lines, that is, along lines of constant $\omega$. The amount of that movement depends on the value of $z_1$ – specifically, the distance of the poles and zeros from the origin is scaled by a factor of $|z|/|z_1|$.

Now think about the case in which $\sigma_1$ is zero, so the factor multiplying the sequence $f(n)$ is $z_1^n = e^{i\omega_1 n}$. In that case, the poles and zeros of $F(z)$ move along lines of constant $\sigma$, which are circles about the origin of the $z$-plane. So the poles and zeros rotate about the origin, and the angle of that rotation is given by the value of $\omega_1$. This rotation in the $z$-plane makes the Z-transform look like this:

$$\mathcal{Z}\left[z_1^n f(n)\right] = F\left(\frac{z}{z_1}\right) = F\left(\frac{e^{(\sigma+i\omega)}}{e^{i\omega_1}}\right) = F(e^{[\sigma+i(\omega-\omega_1)]}), \qquad (5.53)$$

which is a form of frequency modulation, in which the frequency of a time-domain signal is modulated by angular frequency $\omega_1$.

## z-domain Differentiation

If you worked through Section 3.6 you will have seen that differentiation of a Laplace-transform function $F(s)$ in the $s$-domain is related to multiplication of the corresponding time-domain function $f(t)$ by $t$. The discrete-time parallel of that property relates differentiation of the $z$-domain function $F(z)$ and multiplication of the result by $-z$ to multiplication of the discrete time-domain function $f(n)$ by $n$.

The mathematical statement of this Z-transform property is

$$\mathcal{Z}[nf(n)] = -z\frac{dF(z)}{dz}. \qquad (5.54)$$

To see why this is true, start by taking the derivative with respect to $z$ of $F(z)$:

$$\frac{dF(z)}{dz} = \frac{d}{dz}\left[\sum_{n=0}^{\infty} f(n)z^{-n}\right]. \qquad (5.55)$$

Derivation is a linear process, so the derivative can be moved inside the summation, and the discrete-time function $f(n)$ does not depend on $z$, so it moves through the derivative:

$$\frac{dF(z)}{dz} = \sum_{n=0}^{\infty} \frac{d}{dz}\left[f(n)z^{-n}\right] = \sum_{n=0}^{\infty} f(n)\frac{d}{dz}\left[z^{-n}\right]$$

$$= \sum_{n=0}^{\infty} f(n)(-n)z^{-n-1} = -\frac{1}{z}\sum_{n=0}^{\infty} nf(n)z^{-n}. \qquad (5.56)$$

Multiplying both sides of this equation by $-z$ gives

$$-z\frac{dF(z)}{dz} = -z\left[-\frac{1}{z}\sum_{n=0}^{\infty}nf(n)z^{-n}\right] = \sum_{n=0}^{\infty}nf(n)z^{-n} = Z[nf(n)]. \quad (5.57)$$

An intuitive understanding of this property can be gained using the same rationale described in the discussion of the multiplication by $t$ property of the Laplace transform in Section 3.6. In that discussion, the frequency-domain derivative of $F(s)$ (that is, the rate of change of $F(s)$ with $s$) is related to the change in the real time-domain exponential function $e^{-\sigma t}$ and the change in the complex conjugate of the basis function $e^{-i\omega t}$ as $s$ changes, and those changes are proportional to $-tf(t)$.

Applying the same logic to the Z-transform leads to the conclusion that the $z$-domain derivative of $F(z)$ can be related to multiplication of the discrete-time function $f(n)$ by $n$. However, taking the derivative with respect to $z$ of the factor $z^{-n}$ in the Z-transform doesn't just bring down a factor $-n$, it also reduces the power of $z$ by one – that is, it divides $z^{-n}$ by $z$. For that reason, it is necessary to multiply $dF(z)/dz$ by $-z$ in Eq. 5.57 to obtain the Z-transform of $nf(n)$.

## Convolution

The process of convolution is explained in the "Convolution" document on this book's website, and the convolution property of the Laplace transform is described in Section 3.8. A similar property applies to the Z-transform; specifically, this property says that convolution in the time domain is equivalent to multiplication in the $z$-domain.

So if the Z-transform of the discrete time-domain function $f(n)$ is $F(z)$ and the Z-transform of the discrete time-domain function $g(n)$ is $G(z)$, the convolution property can be written as

$$Z[f(n) * g(n)] = F(z)G(z), \quad (5.58)$$

in which $*$ represents convolution.

If both discrete-time sequences $f(n)$ and $g(n)$ have zero amplitude for $n < 0$, the convolution $f(n) * g(n)$ can be written as

$$f(n) * g(n) = \sum_{k=0}^{\infty} f(k)g(n - k) \quad (5.59)$$

and the unilateral Z-transform of $f(n) * g(n)$ is

$$\mathcal{Z}[f(n) * g(n)] = \sum_{n=0}^{\infty} \left[ \sum_{k=0}^{\infty} f(k)g(n-k) \right] z^{-n}. \qquad (5.60)$$

Interchanging the order of summation makes this

$$\mathcal{Z}[f(n) * g(n)] = \sum_{k=0}^{\infty} \sum_{n=0}^{\infty} f(k)g(n-k)z^{-n} = \sum_{k=0}^{\infty} f(k) \sum_{n=0}^{\infty} g(n-k)z^{-n},$$
$$(5.61)$$

in which the function $f(k)$ does not change with $n$ and can therefore be brought outside the summation over $n$.

Now make the change of variable $m = n - k$, or $n = m + k$:

$$\mathcal{Z}[f(n) * g(n)] = \sum_{k=0}^{\infty} f(k) \sum_{m=-k}^{\infty} g(m)z^{-m-k} \qquad (5.62)$$

and since $g(m) = 0$ for $m < 0$, the lower limit of the second summation can be set to zero:

$$\mathcal{Z}[f(n) * g(n)] = \sum_{k=0}^{\infty} f(k)z^{-k} \sum_{m=0}^{\infty} g(m)z^{-m} = F(z)G(z). \qquad (5.63)$$

As in the case of the Laplace transform, it may help your understanding of this Z-transform property to consider the shift-and-multiply process of convolution as a form of polynomial multiplication, in which each term of one function is multiplied by all the terms of the other function. Remembering that the discrete-time sequences $f(n)$ and $g(n)$ are comprised of a series of sampled, orthogonal sinusoids makes it reasonable that only terms with identical frequencies survive the process of multiplication and summation over time, and the Z-transform of those terms is given by the point-by-point multiplication $F(z)G(z)$.

### Initial- and Final-Value Theorems

The initial- and final-value theorems of the Z-transform for sequences $f(n)$ are similar to those of the Laplace transform for time-domain funcitons $f(t)$ discussed in Section 3.9, although there are some notable differences that arise from the discrete-time nature of $f(n)$.

As in the Laplace case, the initial-value theorem relates $f(n)$ to $F(z)$ for small values of time (or sample number) and large values of $z$, while the final-value theorem pertains to large values of $n$ and small values of $z$. These theorems work as long as the limits exist.

For a causal discrete-time sequence $f(n)$ with Z-transform $F(z)$, the mathematical statement of the initial-value theorem is

$$f(0) = \lim_{z \to \infty} F(z) \tag{5.64}$$

and the final-value theorem is

$$\lim_{n \to \infty} f(n) = \lim_{z \to 1} (z-1)F(z). \tag{5.65}$$

To understand why the initial-value theorem is true, it is instructive to write out some of the terms of the Z-transform summation:

$$F(z) = \sum_{n=0}^{\infty} f(n)z^{-n} = f(0)z^0 + f(1)z^{-1} + f(2)z^{-2} + \cdots + f(\infty)z^{-\infty}$$

$$= f(0) + \frac{f(1)}{z} + \frac{f(2)}{z^2} + \cdots + \frac{f(\infty)}{z^\infty}. \tag{5.66}$$

Now taking the limit as $z$ approaches $\infty$ makes this

$$\lim_{z \to \infty} F(z) = f(0) + 0 + 0 + \cdots + 0 = f(0), \tag{5.67}$$

so only the $n = 0$ value of $f(n)$ survives.

The Z-transform final-value theorem take a few more steps to prove, and the most straightforward approach is to begin by taking the Z-transform of $f(n+1) - f(n)$:

$$\mathcal{Z}[f(n+1) - f(n)] = \sum_{n=0}^{\infty} [f(n+1) - f(n)]z^{-n}. \tag{5.68}$$

The linearity of the Z-transform means that

$$\mathcal{Z}[f(n+1) - f(n)] = \mathcal{Z}[f(n+1)] - \mathcal{Z}[f(n)], \tag{5.69}$$

and the Z-transform of $f(n+1)$ can be related to the Z-transform of $f(n)$ by the left-shift relation (Eq. 5.50) with $n_1 = 1$:

$$\mathcal{Z}[f(n+1)] = z^1 F(z) - \sum_{k=0}^{0} f(0)z^{1-0} = z\mathcal{Z}[f(n)] - zf(0). \tag{5.70}$$

Hence Eq. 5.68 can be written as

$$z\mathcal{Z}[f(n)] - zf(0) - \mathcal{Z}[f(n)] = \sum_{n=0}^{\infty}[f(n+1) - f(n)]z^{-n} \qquad (5.71)$$

or

$$(z-1)\mathcal{Z}[f(n)] - zf(0) = \sum_{n=0}^{\infty}[f(n+1) - f(n)]z^{-n}. \qquad (5.72)$$

The next step is to take the limit of both sides of this equation as $z \to 1$:

$$\lim_{z\to1}(z-1)\mathcal{Z}[f(n)] - \lim_{z\to1}zf(0) = \lim_{z\to1}\sum_{n=0}^{\infty}[f(n+1) - f(n)]z^{-n}$$

$$\lim_{z\to1}(z-1)\mathcal{Z}[f(n)] - f(0) = \sum_{n=0}^{\infty}[f(n+1) - f(n)]. \qquad (5.73)$$

Now consider the summation involving the difference between $f(n+1)$ and $f(n)$. As $n$ is incremented from 0 to $\infty$, only the lowest-$n$ and highest-$n$ terms survive the subtraction of $f(n)$ from $f(n+1)$. So in the limit of $n \to \infty$, this summation may be written as $f(n+1) - f(0)$:

$$\lim_{z\to1}(z-1)\mathcal{Z}[f(n)] - f(0) = \lim_{n\to\infty}f(n+1) - f(0) \qquad (5.74)$$

and canceling the $f(0)$ terms gives

$$\lim_{z\to1}(z-1)\mathcal{Z}[f(n)] = \lim_{n\to\infty}f(n), \qquad (5.75)$$

in accordance with Eq. 5.65.

The concepts in this chapter may seem quite basic, but there is nothing like working problems to make sure that your understanding of this material is deep enough to allow you to tackle the next level of books and papers dealing with Fourier, Laplace, and Z-transforms. The problems in the following section are designed to help you do that, so you should take the time to work through them – and remember that full, interactive solutions to every problem are freely available on this book's website.

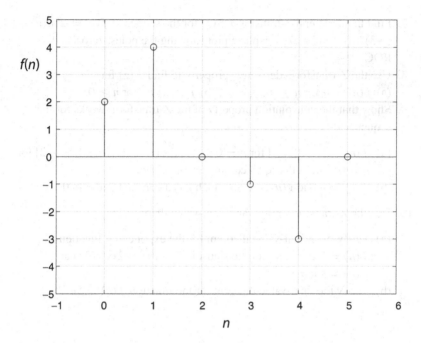

## 5.4 Problems

1. Find the Z-transform $F(z)$ of the sequence shown above, then use the right-shift version of the time-shift property of the Z-transform to find $F(z)$ for this sequence, shifted two samples to the right.

2. Use the inverse Euler relation and the approach shown in Section 5.2 to verify Eq. 5.43 for $F(z)$ for the discrete-time sine function $f(n) = \sin(\omega_1 n)$.

3. Use the Z-transform examples of Section 5.2 and the linearity property discussed in Section 5.3 to find the Z-transform $F(z)$ of the sequence $f(n) = 5(2^n) - 3e^{-4n} + 2\cos(6n)$.

4. Use the definition of the unilateral Z-transform (Eq. 5.9) to find $F(z)$ for $f(n) = \delta(n - k)$ and compare your result to the result of using the shift property (Eq. 5.49).

5. Use the approach shown in the "Time-Shifting" subsection of Section 5.3 to confirm the left-shift relation (Eq. 5.50), then apply that relation to the sequence of Problem 1, shifted two samples to the left.

6. Use the inverse Euler relation for the cosine function and the multiply-by-an-exponential property along with the Z-transform of the unit-step function $u(n)$ to find $F(z)$ for $f(n) = A^n \cos(\omega_1 n) u(n)$.

7. Find the unilateral Z-transform $F(z)$ for the sequence $f(n) = 2^n u(n) + (-3)^n u(n)$ and make a $z$-plane plot showing the poles, zeros, and ROC.

8. Use the Z-transform derivative property to find $F(z)$ for
   (a) $f(n) = n$ for $n \geq 0$        (b) $f(n) = n^2$ for $n \geq 0$.

9. Show that the convolution property of the Z-transform works for the sequences

   (a) $f(n) = [-1, 0, 4, 2]$ for $n = 0$ to $3$ and $g(n) = [3, -1, 1, 5, -2]$ for $n = 0$ to $4$; both sequences are zero elsewhere.
   (b) $f(n) = n$ and $g(n) = c$, in which $c$ is a constant and $n \geq 0$.

10. Use the Z-transform inital- and final-value theorems to

   (a) Verify the initial-value theorem for the exponential function $f(n) = -e^{-2n}$ and for the sinusoidals $f(n) = 2 \cos(3n)$ and $f(n) = \sin(n)$.
   (b) Verify the final-value theorem for the function $f(n) = 5e^{-3n}$.

# Further Reading

Brigham, E., *The FFT*, Prentice-Hall 1988.

Dyke, P., *An Introduction to Laplace Transforms and Fourier Series*, Springer 2014.

El-Hewie, M., *Laplace Transforms*, Mohamed El-Hewie 2013.

Fleisch, D., *A Student's Guide to Vectors and Tensors*, Cambridge University Press 2011.

Fleisch, D. and L. Kinnman, *A Student's Guide to Waves*, Cambridge University Press 2015.

Graf, U., *Applied Laplace Transforms and Z-Transforms for Scientists and Engineers*, Springer Basel AG 2004.

James, J. *A Student's Guide to Fourier Transforms*, Cambridge University Press 2011.

Mix, D., *Fourier, Laplace, and Z-Transforms*, Dwight Mix 2019.

Oppenheim, A. and R. Schafer, *Discrete-Time Signal Processing*, Prentice-Hall 1989.

Riley, K., M. Hobson, and S. Bence, *Mathematical Methods for Physics and Engineering*, Cambridge University Press 2006.

Spiegel, M., *Laplace Transforms*, Schaum's Outline Series, McGraw-Hill 1965.

# Index

Printed in the United States
by Baker & Taylor Publisher Services

Printed in the United States
by Baker & Taylor Publisher Services